Mathematical Marvels

Adventures in PROBLEM SOLVING

Shailesh Shirali

Universities Press

Adventures in Problem Solving

UNIVERSITIES PRESS (INDIA) PRIVATE LIMITED

Registered Office
3-6-747/1/A and 3-6-754/1, Himayatnagar, Hyderabad 500 029, Telangana, India
info@universitiespress.com; www.universitiespress.com

Distributed by
Orient Blackswan Private Limited

Registered Office
3-6-752, Himayatnagar, Hyderabad 500 029, Telangana, India

Other Offices
Bengaluru, Bhopal, Chennai, Guwahati, Hyderabad, Jaipur, Kolkata, Lucknow, Mumbai, New Delhi, Noida, Patna, Visakhapatnam

First Published 2002
Reprinted 2008, 2009,2017, 2019

501533

ISBN 978-81-7371-413-9

Typeset by
OSDATA, Hyderabad 500 029

Printed at
Graphica Printers & Binders, Hyderabad 500 013

Published by
Universities Press (India) Private Limited
3-6-747/1/A and 3-6-754/1, Himayatnagar,
Hyderabad 500 029, Telangana, India

Contents

Preface

What is "problem solving"? The great mathematician David Hilbert said that, in essence, problems are the life blood of mathematics, and Paul Halmos added a rider: that the complementary activity – theory building – provides the soul of the subject. Here is Halmos' view:

> Mathematicians sometimes classify themselves as either problem-solvers or theory-creators. The problem-solvers answer yes-or-no questions and discuss vital special casesand concrete examples that are the flesh and blood of all mathematics; the theory-creators fit the results into a framework, illuminate it all, and point it in a definite direction—they provide the skeleton and the soul of mathematics. One can be both a problem-solver and a theory-creator, but [is] usually one or the other. The problem-solvers make geometric constructions, the theory-creators discuss the foundations of Euclidean geometry; the problem-solvers find out what makesswitching diagrams tick, the theory-creators prove representations for Boolean algebras.

At the research level, good problems often play a vital role by directing the future course of a subject. When Hilbert offered his famous list of twenty-three problems at the International Congress of Mathematicians in 1900, he must surely have had this in mind; but perhaps even he would have been surprised at the extent to which his list shaped the course of research in mathematics in the 20th century. It is not hard to see why this would happen; a concerted attack on a difficult problem inevitably throws up a large number of challenges—new notation has to be devised, new concepts introduced, new connections found; and often this work provides the foundation of a new branch of mathematics. This phenomenon can be seen many times in the history of the subject. Consider, for instance, the role played by the Königsberg bridges problem (posed and solved by Euler) in the origins of graph theory, or the growth of this subject achieved through work on the four-colour problem; or the progress achieved in number theory as a result of the work on Fermat's 'last theorem'; or the discovery of non-Euclidean geometries by Lobachevsky and Bolyai (independently) as a result of their efforts to prove the fifth postulate of Euclid (the "parallel postulate"); or the growth of recursive set theory through the efforts to solve Hilbert's tenth problem (on algorithms to solve Diophantine equations). More such instances can readily be quoted. As for pedagogy, the benefits of problem solving will be attested to by practically every teacher of mathematics. In some countries, notably Hungary, Russia and Romania, problem solving and its dual activity, that of problem posing, have been developed into a fine craft and have entered the teaching of mathematics in a big way. The sizeable growth of interest in the Mathematics Olympiads over the last two decades attests to the increasing awareness of the importance of problem solving; problem solving is being increasingly talked about in educational circles these days.

This book is written for problem buffs, but it is not a book about strategies or heuristics. There are many excellent books on the art of problem solving already written; the best amongst them (Polya's two books, *How To Solve It* and *Mathematical Discovery*) would be difficult to better. The present work is more in the spirit of Newman's *A Problem Seminar* and Polya and Szëgo's *Problems and Theorems in Analysis*, both of which offer problem sequences rather than isolated problems. When I first started writing this book, the title I had in mind was *The Art of Problem Solving*; but once the book got under way, it became clear that an alternative title was needed. I briefly considered *Episodes in Problem Solving* before finally opting for the present title.

The first chapter discusses the nature of problem solving, but this is the only place in the book where strategies per se are discussed. Following this we take up different topics and solve numerous problems—indices and surds, logarithms, Euclidean geometry, algebraic and geometric inequalities, grids and chess boards, cryptarithms (arithmetic problems in disguise), magic squares, logic, combinatorics, number theory, coin weighing and two-person games (e.g., Nim). As will be seen, the coverage is wide, and the number of problems posed is very large. Some of them are quite difficult, and occasionally an indication has been made of this (the symbol '$\boxed{*}$' is affixed to the problem). The problems themselves are for the most part from published sources (e.g., Smullyan's *What is the Name of this Book?*, Martin Gardner's *Mathematical Diversions*, *Further Mathematical Diversions*, and others of that extended family), and are thus not original.

Most of all, the book is a celebration of mathematics through problems. It is written for those who enjoy doing mathematics; may their tribe increase, and may it increase their enjoyment.

Acknowledgements Many of my most memorable problem solving adventures occurred during the selection and training camps for the IMOs that my colleagues C R Pranesachar, B J Venkatachala, C S Yogananda and I conducted on many occasions, and I wish to thank them. Thanks are also due to the National Board of Higher Mathematics (TIFR, Mumbai), this being the apex body behind Olympiad activity in India.

My initiation into the Olympiads came from the late Prof S Izhar Husain of Aligarh Muslim University, from Prof A M Vaidya of Gujarat University, from Prof Phoolan Prasad of the Indian Institute of Science, and from Dr R Subramanian (formerly of BARC), and I thank them too.

Chapter 1

Introduction

1.1 The nature of problem solving

Is problem solving an art or a science?

It is certainly not purely a science—if it were, then problem solving would by and by become a routine affair, and over the course of time, techniques would become available to solve all possible types of problems. Research would proceed at a brisk pace and unsolved problems would rapidly be dealt the *coup de grâce*. Allowing the imagination to run wild, one can fantasize about the effects of nationalism on the progress of mathematics. Competent bureaucrats would then issue crisp directives on behalf of the government:

> *Tenders are to be issued for settling the following four long-standing disputes:*
>
> 1. *Is the Riemann hypothesis true?*
> 2. *Is* P = NP*?*
> 3. *Is there always a prime between* n^2 *and* $(n+1)^2$ *for positive integers* n*?*
> 4. *Does there exist a good primality testing algorithm?*
>
> *Suitable personnel are to be appointed. Tasks are to be completed within 2 years of commencement of the project, or else heads will roll; . . . .*

Unfortunately (or fortunately) however, research does not quite proceed in this manner! On the other hand, it *is* part of the mathematician's trade to be on the lookout for techniques that make the solving of particular kinds of problems a routine affair.

For example, consider the problem of solving a quadratic equation. Today, this is considered routine; there is a simple formula available which can be programmed into a calculator. So the calculator can, in a sense, "solve" this problem. Likewise, the solution of a linear equation with one unknown, for example, $2x + 5 = 11$ (the unknown being x), is routine, as is the solution of a pair of linear equations (better known as 'simultaneous equations') in two variables, for example,

$$\left.\begin{array}{rcrcr} 2x & + & 5y & = & 11 \\ x & + & 3y & = & 6 \end{array}\right\} \text{ (with unknowns } x \text{ and } y\text{).}$$

Not all readers may be familiar with the fact that solving single-variable cubic and quartic equations, for example,

$$x^3 = x + 10, \quad \text{or} \quad x^4 = x + 100,$$

(the unknown is x in each case) is considered routine too. The solutions of the general cubic and quartic equations represent great chapters in the history of mathematics, occurring in the middle of the 1500s, and the mathematicians involved all being from Italy.[1] Thus, cubic and quartic equations are no longer to be viewed as problems, in the sense that they have (in principle) been solved completely.

However, not all problems are of this nature. Consider the following:

> **Problem**
> *Find all 4-digit numbers N for which the number N′ obtained by arranging the digits of N in reverse order is 9 times the original number; i.e., $N' = 9 \times N$. (We work throughout in decimal notation, i.e., in base-10.)*

Here, we have no ready formula at hand. A hit-and-miss approach can be tried: *Is 1009 the answer?*—no, because $1009 \times 9 = 9081$, which is not the reversal of 1009. *Is 1019 the answer?*—no, because $1019 \times 9 = 9171$, which is not the reversal of 1019; . . . . But this is

[1]The mathematicians are Cardano, Tartaglia and Ferrar. The human aspect of the story is of great interest, but we shall not go into it here. An excellent account of the story is given in William Dunham's wonderfully readable book, *Journey Through Genius.*

too tedious!—not to say unenlightening. The use of a computer will quicken the search very greatly, and we need have no fears about boredom for the computer! In fact, many people use the computer for just this type of work – for conducting hit-and-miss searches – particularly when the problem at hand is complex and the mathematician finds it hard to acquire a 'feel' for the problem. The computer thus represents a powerful tool in problem solving, and one must be prepared to make use of it whenever appropriate. This is more true now than ever before, with the easy availability of many excellent software packages (Derive, Mathematica, Maple, ...), which are useful not only for numerical and graphical work, but also for symbolic work. Such packages can be of immense help in building up a database of facts about the problem—an experience-bank, as it were. Ultimately, intuition rests on precisely such a database.

We shall discuss the number-reversal problem later on in this chapter.

1.2 Mathematics on a computer

Here is a simple example of a problem where a computer greatly helps in the search for a solution.

Problem
Consider the function $F(x) = 3x^4 + 4x^3$. Show that the number $F(F(F(\ldots F(9))))$, with F repeated ten times, acting on an initial input of 9, has more than one thousand 9s when expressed in decimal (base-10) notation.

The numbers involved here are so large that one quickly feels a sense of hopelessness! A workout on the computer yields the following:

$$\begin{aligned}
F(9) &= 22599 \\
F(F(9)) &= F(22599) \\
&= 782534990456559999 \\
F(F(F(9))) &= F(F(22599)) \\
&= F(782534990456559999) \\
&= 1124\ldots\ldots 54201599999999.
\end{aligned}$$

($F(F(F(9)))$ is a number with 73 digits!) A discovery is made almost immediately: $F(9)$ has two 9s at its units end, $F(F(9))$ has four 9s at its units end, $F(F(F(9)))$ has eight 9s at its units end, ... The pattern cannot be missed! Therefore, the hypothesis is:

> *The number $X(n)$ obtained by n applications of F to an input of 9 has 2^n nines at its units end;*

or, still more compactly,

> *The number $X(n)+1$ is divisible by 10^{2^n}*

Having framed the hypothesis, an easy proof can be found based on induction. The problem has thus been solved; well, almost!

⋆ ⋆ ⋆

Here is a full proof. The method we use is *induction*: we check that the result is true for $n = 1$ (it is, as noted above); then we assume it to be true for $n = 1, 2, 3, \ldots, N$ and proceed to show that is also true for $n = N + 1$. If this attempt is successful, then the principle of induction assures us that the result is true for all positive integers n. For convenience, we write y for $X(N)$ and z for $X(N+1)$. By assumption, $y+1$ is divisible by 10^{2^N}. Let $y+1 = 10^{2^N}k$. Then, $y = 10^{2^N}k - 1$ and

$$z = F(y) = 3\left(10^{2^N}k - 1\right)^4 - 4\left(10^{2^N}k - 1\right)^3$$

Expanding the two bracketed expressions, we easily find that

$$z = 10^{2^{N+1}}k' - 1$$

for some integer k', so $z+1$ is divisible by $10^{2^{N+1}}$. This establishes the result.

1.3 Triangle inequality

Here is an example from the author's own experience. We shall consider the following problem:

> **Problem**
> *Let $\triangle ABC$ with sides a, b and c have perimeter 2. Prove that $a^2 + b^2 + c^2 + 2abc < 2$.*

I substituted $c = 2 - a - b$ into the left side of the inequality and then attempted to obtain, using calculus, the maximum possible value of the resulting expression, namely:

$$a^2 + b^2 + (2-a-b)^2 + 2ab(2-a-b).$$

This turned out to be excessively tedious! After several fruitless attempts, I began to sense that the difficulty lay in the fact that the inequality was non-homogeneous (it has terms such as $2abc$ and a^2, which have different degrees). Could I somehow "homogenize" the inequality? I found that I could: by exploiting the fact that $a+b+c=2$, the following two equalities then become available:

$$a^2+b^2+c^2 = \left(\frac{a+b+c}{2}\right)\left(a^2+b^2+c^2\right),$$

$$2 = \frac{(a+b+c)^3}{4}.$$

Proceeding along such lines, I framed the following problem:

Problem
Show that for a triangle with sides a, b and c, the following inequality holds:

$$\frac{(a^2+b^2+c^2)(a+b+c)}{2} + 2abc - \frac{(a+b+c)^3}{4} < 0.$$

The inequality is now homogeneous: all its terms are of degree 3.

Using Derive, I found a quick solution to the problem. I first multiplied the left side of the above expression by 4 to clear the fractions and obtained the expression:

$$2(a^2+b^2+c^2)(a+b+c) + 8abc - (a+b+c)^3.$$

I then got Derive to open this expression and collect similar terms together, using the `Simplify` command. This was quickly done and the result was the following expression:

$$a^3 - a^2(b+c) - a(b^2-2bc+c^2) + (b+c)(b^2-2bc+c^2),$$

which is visibly divisible by $a-b-c$; to see why, club the first two terms together and then the last two terms. Now, symmetry could be invoked: since the expression

$$2(a^2+b^2+c^2)(a+b+c) + 8abc - (a+b+c)^3$$

is symmetric in a, b, c; if it is divisible by $a-b-c$, then it must be divisible by $b-c-a$ and by $c-a-b$ as well. So I got Derive to divide the expression by the quantity $(a-b-c)(b-c-a)(c-a-b)$. The quotient was returned immediately: 1! Thus, the full factorization of the expression has been obtained; the expression is *equal* to $(a-b-c)(b-c-a)(c-a-b)$.

The problem now acquired the following equivalent form:

Problem
Let a, b, c be the sides of a triangle. Prove that the quantity $(a-b-c)(b-c-a)(c-a-b)$ is negative.

But this is obvious! According to the triangle inequality (*"the sum of the lengths of any two sides of a triangle exceeds the length of the third side"*), the quantities $a-b-c$, $b-c-a$ and $c-a-b$ are all negative; therefore, their product too is negative. The problem has thus been solved!

There are other methods for solving this problem, but the manner in which the computer aided the solution is striking and worthy of note.

1.4 Riffle shuffle

We present another problem where the computer helps in gaining insight into a problem. This problem concerns the *riffle shuffle*, which players often use to shuffle a pack of cards. The reader may have seen the shuffle being done during a game of bridge or poker (possibly on the TV screen!). When done with skill and grace, it looks most impressive.

The shuffle is done as follows. Let the pack have n cards, where n is an even number. The pack is 'cut' at its middle point into two packs, each with $n/2$ cards. The two packs are now held side by side in the hands, face downwards, and the cards are dropped one by one, alternately from the two packs, starting from the bottom, till both packs are exhausted. Following this, the cards are gathered up. This constitutes one round of the riffle shuffle.

We illustrate what happens with a pack containing 6 cards. Let the cards be numbered $(1, 2, 3, 4, 5, 6)$, starting from the top and proceeding downwards. After the first 'cut' the two halves are placed side by side, as shown below.

1	4
2	5
3	6

The cards are now dropped from the hands: first #6, then #3, then #5, then #2, then #4 and, finally, #1. The result is that the cards are now in the following order: $(1, 4, 2, 5, 3, 6)$.

What has been done once can always be done again, and we shall proceed to do so. Cutting the pack into two halves and placing the two halves side by side, we obtain the following:

1	5
4	3
2	6

Dropping the cards and gathering them as described, we find that the cards are now in the order $(1, 5, 4, 3, 2, 6)$. We then go through the procedure yet again; the cards now get arranged in the order $(1, 3, 5, 2, 4, 6)$. A fourth repetition of the shuffle brings the cards into the order $(1, 2, 3, 4, 5, 6)$; that is, the original order has been restored and it has taken *four* shuffles for this to happen.

If we work with 8 cards, a similar thing happens. The order of the cards at the intermediate stages is displayed below.

Stage 0	$(1, 2, 3, 4, 5, 6, 7, 8)$
Stage 1	$(1, 5, 2, 6, 3, 7, 4, 8)$
Stage 2	$(1, 3, 5, 7, 2, 4, 6, 8)$
Stage 3	$(1, 2, 3, 4, 5, 6, 7, 8)$

Thus, in this case, *three* repetitions of the shuffle have sufficed to restore the original order of the cards.

We begin to suspect at this point that no matter what value n has, the original order will always be restored after a certain number of repetitions of the shuffle. This is so, but to see why demands a careful reflection on what really happens in a shuffle. In fact, the phenomenon takes place for *any* shuffle: if it is done over and over again, the starting configuration will inevitably be reached at some point in time.

(*Note. Here, by 'shuffle' we mean any fixed permutation of the cards.*)

We now define a function r_n for even numbers n, as follows. Let the card pack have n cards; then r_n is the least number of repetitions of the riffle shuffle needed to reach the original configuration (so $r_6 = 4$, $r_8 = 3$, $\ldots$). How does r_n change with n? To make rapid progress, we use the computer to generate data. A sample Mathematica program that helps in computing r_n, for a given n, is displayed below.

```
r[n_]:= (f[i_]:= If[Mod[i,2]==1, (i+1)/2, (i+n)/2];
```

```
SetAttributes[f, Listable]; k = 1; a = f[2];
While[Not[a==2], (a = f[a]; k = k+1)]; k)
```

Running the program for some sample values of n, we find the following:

n	2	4	6	8	10	12	14	16	18	20
r_n	1	2	4	3	6	10	12	4	8	18

n	22	24	26	28	30	32	34	36	38
r_n	6	11	20	18	28	5	10	12	36

What do these figures suggest? Here is one observation: *If $n = 2^k$ for some positive integer k, then $r_n = k$.*

EXAMPLE. $r_4 = 2, r_8 = 3, r_{16} = 4, r_{32} = 5, \ldots$.

This is a pretty discovery indeed! In fact, it is very typical of a discovery that a computer can help us make.

There is plenty of material in this problem for investigation and further study, and we shall not spoil the reader's fun by giving away the answers at this stage. Instead, we close the section with a challenge: *Find the manner in which r_n depends on n.*

1.5 Number reversal

We return to our $N' = 9 \times N$ problem. A computer program may be written to yield a solution, based on the following "brute-force" search algorithm.

1. Start with the value $n = 1000$.
2. Compute n', the 'reversal' of n.
3. Check if $n' = 9n$.
4. If 'yes', print out the value of n.
5. If $n = 9999$ then go to step 7.
6. If $n < 9999$, then $n \to n + 1$; go to step 2.
7. Print "Search completed."
8. Stop.

COMMENT. Here "$n \to n+1$" means "increment n by 1"; equivalently, "replace n by $n+1$". The only tricky part is the computation of n' (the reversal of n). How do you get the computer to do this task?

It is obvious that the program will give all possible answers to the problem. If all we want is such a list, then we do not need to do anything further. But what if we want more? What if we wish to understand the "why" of the answer? Surely this is what a mathematician really wants: *an understanding of the problem itself, not merely a statement of the answer.* If this is our desire, then we are thrown back on our own resources. We now have to work entirely on our own, and see for ourselves what the problem is all about. We shall proceed to do just this.

Notation The 2-digit base-10 number with digits a, b will be written as $\overline{\mathsf{ab}}$ and not as ab (which suggests the product $a \times b$). The line above the expression and the difference in font serve to draw attention to the fact that we have altered the 'mode' and are considering the digital representation of the number on hand.

⋆ ⋆ ⋆

Let the digits of the number, reading from left to right, be A, B, C and D, respectively (D is the units digit, C is the tens digit, and so on). The number itself is written as $\overline{\mathsf{ABCD}}$. We write the given property as follows:

$$\overline{\mathsf{ABCD}} \times 9 = \overline{\mathsf{DCBA}},$$

or in long multiplication form as:

$$\begin{array}{cccc} \mathsf{A} & \mathsf{B} & \mathsf{C} & \mathsf{D} \\ & & \times & 9 \\ \hline \mathsf{D} & \mathsf{C} & \mathsf{B} & \mathsf{A} \\ \hline \end{array}$$

What could A be? If A were greater than 1, i.e., $\mathsf{A} \geq 2$, then we would have $\overline{\mathsf{ABCD}} \geq 2000$; but $9 \times 2000 = 18000$, a 5-digit number, whereas the product, $\overline{\mathsf{DCBA}}$, is a 4-digit number. So A cannot exceed 1, and it follows that $\mathsf{A} = 1$.

What could D be? Arguing from the units end, $9 \times \mathsf{D}$ is either A or an integer with units digit A. Since $\mathsf{A} = 1$, it follows that $9 \times \mathsf{D}$ is a number with units digit 1. This forces D to be 9 (because $9 \times 9 = 81$).

We could also have deduced that $\mathsf{D} = 9$ by arguing from the thousands digit; do you see how?

We now collect together what we have uncovered:

$$\begin{array}{cccc} 1 & \mathsf{B} & \mathsf{C} & 9 \\ & & \times & 9 \\ \hline 9 & \mathsf{C} & \mathsf{B} & 1. \\ \hline \end{array}$$

There are now two ways of proceeding. One is by noting that $1200 \times 9 = 10800$, a 5-digit number. This implies that $\mathsf{B} \le 1$, so $\mathsf{B} = 0$ or 1. Arguing from the tens column, we have a carryover of 8 from the units column, so $(9 \times \mathsf{C}) + 8$ has B for its last digit. We shall try out both possibilities for B.

- If $\mathsf{B} = 1$, then $9 \times \mathsf{C} + 8$ has units digit 1, so $9 \times \mathsf{C}$ has units digit 3; therefore, $\mathsf{C} = 7$ and $\overline{\mathsf{ABCD}} = 1179$. But this doesn't work: $1179 \times 9 \neq 9711$. So the assumption that $\mathsf{B} = 1$ does not lead to a solution. In mathematical terms, the assumption is 'self-contradictory'.

- Next we try out the possibility: $\mathsf{B} = 0$:

$$\begin{array}{cccc} 1 & 0 & \mathsf{C} & 9 \\ & & \times & 9 \\ \hline 9 & \mathsf{C} & 0 & 1. \\ \hline \end{array}$$

  Since $(9 \times \mathsf{C}) + 8$ is a number with units digit 0, $9 \times \mathsf{C}$ is a number with units digit 2; therefore, $\mathsf{C} = 8$ and $\overline{\mathsf{ABCD}} = 1089$.

Does this work? It does: $1089 \times 9 = 9801$. So we have not only obtained an answer, we have shown that it is the *only* answer. How much more satisfying this is than a computer-generated output!

⋆ ⋆ ⋆

Let us study this particular example once more. Suppose that the problem is modified to the following:

Problem
Find all 4-digit numbers N for which the number N' obtained by reversing the digits of N is 7 times the original number.

As earlier, we write $N = \overline{\mathsf{ABCD}}$:

$$\begin{array}{cccc} \mathsf{A} & \mathsf{B} & \mathsf{C} & \mathsf{D} \\ & & \times & 7 \\ \hline \mathsf{D} & \mathsf{C} & \mathsf{B} & \mathsf{A} \\ \hline \end{array}$$

Let us consider what might be the value of A. Since 2000×7 is a 5-digit number, A cannot exceed 1. It cannot be 0, as it is the leading digit of N; $\therefore \mathsf{A} = 1$. Therefore (arguing from the units end), $7 \times \mathsf{D}$ is a number with units digit 1, and we infer that $\mathsf{D} = 3$. Thus, we have:

$$\begin{array}{rrrr} 1 & \mathsf{B} & \mathsf{C} & 3 \\ & & \times & 7 \\ \hline 3 & \mathsf{C} & \mathsf{B} & 1. \\ \hline \end{array}$$

However, this cannot possibly work out for *any* choice of B, C, because the first digit of the product would have to be at least 7, whereas we have just found that it is 3.

We conclude that *such numbers do not exist*. In other words, we reach a negative conclusion to the problem.

1.6 General comments

The reader may feel some disappointment, maybe even discomfort, at the negative conclusion above. In fact, there is cause for elation, not disappointment—we have found something new! This is quite common in mathematics; it frequently happens that a problem as originally stated has no solution at all. The history of mathematics shows that such situations have recurred time and again: problems are posed, exhaustively analyzed and, in the end, found to have no solution. Of course, in the process a rich and pleasing understanding of the problem emerges. For instance, the famous geometric construction problems[2] considered by the Greeks more than two thousand years ago:

(a) the trisection of an arbitrary angle;

(b) the squaring of a circle;

(c) the duplication of a cube;

all ultimately turned out to be unsolvable. That is, we have the following 'negative' results:

> *There is no ruler-and-compass construction procedure whereby: (a) an arbitrary angle can be trisected, (b) a square can be constructed with area equal to that of a given circle, and (c) a line segment can be constructed having a length of $\sqrt[3]{2}$, the unit of length being that of a given line segment.*

[2]The only instruments allowed are a pair of compasses, an unmarked ruler, and (of course!) a pencil and an eraser.

The discovery that these problems have no solution is itself part of mathematics! Proving these negative statements is no mean feat and involves a lot of very exciting mathematics.

⋆ ⋆ ⋆

Returning to the problems considered above ($N' = 9N$ and $N' = 7N$), note the manner in which we moved back-and-forth between the units digit and the thousands digit. This kind of reasoning is often used in the analysis of such problems ("cryptarithms"). Note also how we moulded our approach to suit the problem. Thus, our approaches in solving the two problems

$$\overline{\mathsf{ABCD}} \times 9 = \overline{\mathsf{DCBA}},$$
$$\overline{\mathsf{ABCD}} \times 7 = \overline{\mathsf{DCBA}},$$

were similar, but not identical. This means that we were not using a fixed formula or algorithm, as it were—a fixed approach which, when mechanically carried out, would provide us with all possible answers. On the contrary, we were ready at each point to grab hold of any particular features that the problem had to offer and to make full use of them.

This is precisely the art of problem solving! There is no fixed formula or precisely defined set of rules that would enable us to solve every problem; that is, there is no *algorithm* available for solving every possible problem. Rather, we have approaches based on certain general considerations or principles, coupled with an awareness of the finer details of the particular problem at hand.

⋆ ⋆ ⋆

It should be clear now that problem solving is not a science at all, but an art. Rather, one can say the following: *the attempt to convert problem solving into a science is itself part of the scientific endeavour.* In a sense, this describes the work of George Polya (and also the work being conducted in artificial intelligence).

For mathematics itself, problem solving constitutes its life blood. This is no exaggeration: vast amounts of mathematics have been created in the attempt to solve difficult problems. One only has to look at the history of Fermat's Last Theorem to see the truth of this; much of modern Algebraic Number Theory and Class Field Theory are by-products of the efforts to prove Fermat's assertion, made some three-and-a-half centuries ago.[3] The story behind the proof is a most exciting one, with an imposing list of mathematicians taking part (Fermat, Euler, Legendre, Gauss, Lamé, Dirichlet, Kummer, Kronecker, Mordell, Faltings, Weil, Serre, Wiles, . . . ; whew!), and has been documented in several places.

Likewise, the enormous amount of energy that went into trying to prove the infamous Parallel Postulate (the fifth and last in Euclid's list) eventually produced a gem: non-Euclidean Geometry.

In fact, it is not necessary to look to advanced mathematics to experience at first hand the value of problem solving. Work hard at any difficult problem and examine, after it is all over, how much one has learnt in the process (even if the attempt has been unsuccessful). It is inevitable that this will happen, because in tackling a really difficult problem one has taken on a formidable adversary and one is forced to use every available weapon. It is a battle in which one's abilities are stretched to the utmost. And in the process, one learns . . . .

In this connection, a remark about *perseverance* is appropriate. Implicit in many students' attitudes towards problem solving is the assumption that it should be possible to solve any problem within, say, ten minutes or half-an-hour at the most. If this does not turn out to be the case – i.e., if one has not 'gotten anywhere' at the end of the stipulated time period – then the conclusion is, "There must be some trick involved, it is too difficult, I'm just not getting it, and anyway it is probably not worth the effort"; or, "This is just not my subject"; or other conclusions resembling these in tone and spirit. As the reader will appreciate, the effect of this type of assumption is rather destructive and needs to examined critically.

⋆ ⋆ ⋆

Earlier, we had posed the question: *"Is problem solving an art or a science?"* and suggested that it resembles an art more than a science. However, it would not be satisfactory to leave the matter

[3]For the sake of correctness, I should add that not all of algebraic number theory has come from efforts to prove Fermat's Last Theorem. There is another result in number theory known as the Law of Quadratic Reciprocity, discovered empirically by Euler and Legendre and later proved by Gauss, who called it the 'golden theorem' or *theorema aureum* of mathematics. Efforts to find generalizations of the theorem, that is, to find the laws of cubic and quartic reciprocity, were responsible for tremendous advances in the field of algebraic number theory. But this amended statement only goes to further prove the point.

at this stage and move on. The point is that there are approaches to problem solving that do help. The famous mathematician-educator George Polya has done an enormous amount of work on developing heuristics for problem solving. A 'heuristic' is simply a guideline or a rough-and-ready rule to solve a particular type of problem; practicing engineers, for example, rely a great deal on them. Heuristics are discussed in detail in Polya's two books, *Mathematical Discovery* and *How to Solve It*.

For instance, a very obvious first step in tackling a problem in geometry is the drawing of a good diagram, possibly using colour, dotted lines, and solid lines, to distinguish between different parts of the diagram. Obviously this is not going to solve the problem for us—but it helps! (It is curious to note how often this elementary step is ignored.)

Here are a few general principles that do help.

- Try out particular cases; restrict the problem in various ways and check out the results; check extreme cases. (There is an art to this too! One may end up restricting the problem to such a degree that the special case offers no insight into the general situation. For instance, faced with a problem about an arbitrary quadrilateral, if we start by solving it for the special case of an isosceles trapezium, we are not likely to be any closer to the solution of the original problem. I am quoting from an actual instance here!—it occurred in the International Mathematical Olympiad of 1989, held that year in Braunschweig, West Germany.)

- Generate number sequences associated with the problem; examine the sequences for patterns; make hypotheses; test them out.

- Break down the problem; find the step that might imply the final result to be proved; see if the problem can be divided into stages ('modules') that are somewhat independent of one other.

- Loosen the problem; remove some of the restrictions and examine the effect of the removal; then restore the restriction and see what happens.

We now present some case-studies that serve to illustrate these principles.

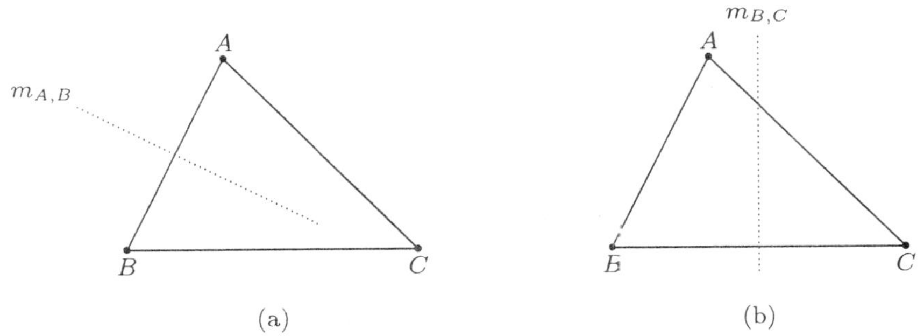

Figure 1.1. *Finding a point O such that $OA = OB = OC$*

1.7 Points on a plane

> **Problem**
> *Given three distinct coplanar points A, B and C, find a point O such that $OA = OB = OC$. Is it possible for there to be more than one such point, or no such point at all?*

We first loosen the problem a bit: we replace '$OA = OB = OC$' by '$OA = OB$'. Clearly, there are infinitely many points O with this property, namely, the points on the line labelled m_{AB} in Figure 1.1(a). This line is the perpendicular bisector or *mediator* of segment AB. If A, B are points on a sheet of paper, then m_{AB} can be constructed by folding the paper so that A falls upon B.

Instead of loosening the requirement to '$OA = OB$', we could just as well have opted for '$OB = OC$'. This property holds for all points on m_{BC}, the mediator of segment BC, as shown in Figure 1.1(b).

If these two lines intersect at some point O, then we have $OA = OB$ as well as $OB = OC$. Therefore, $OA = OB = OC$, and we have found our required point (Figure 1.2).

It is clear from the preceding analysis that there cannot be more than one point, O, with the given property; for m_{AB} and m_{BC}, being straight lines, cannot meet at more than one point. Is it possible that the lines do not meet at all? Yes—if they are parallel! This will happen if AB and BC are part of the same line, i.e., if A, B, C are collinear. In this case, no point O with the required property exists (Figure 1.3).

Our problem has thus been solved fully. *A unique point O with the required property exists in all instances when A, B and C are not collinear; and when the points are collinear, no such point exists.* As

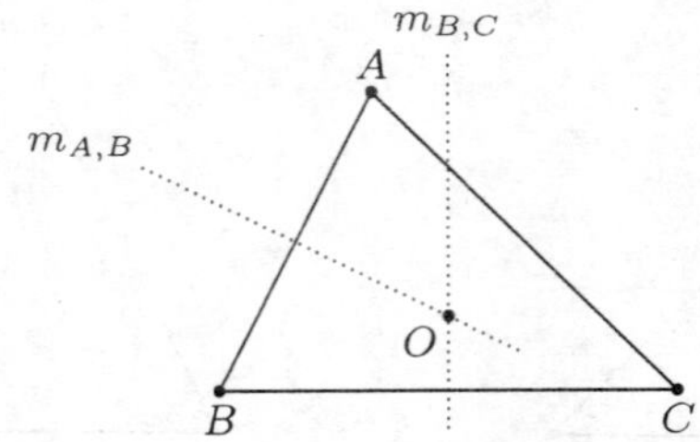

Figure 1.2. *Finding a point O such that $OA = OB = OC$.*

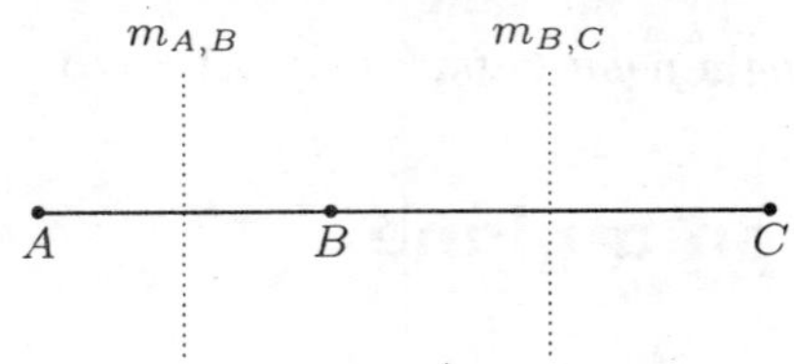

Figure 1.3. *What if A, B, C are collinear?*

a corollary we have the theorem: *The mediators of the three sides of a triangle meet in a point.*

The approach described above can be used for many problems of this nature, for instance:

1. Given three points A, B, C, find all points P with the property that the areas of triangles PAB, PBC, PCA are equal.
2. Given three points A, B, C, find all points P such that the altitudes through P of the triangles PAB, PBC, PCA are equal in length.
3. Given three points A, B, C, find all points P such that $\angle APB = \angle BPC = \angle CPA$.
4. Given three points A, B, C, find all points P such that $\angle PAB = \angle PBC = \angle PCA$.
5. Given three points A, B, C, find all points P such that $PA : PB : PC = 1 : 2 : 3$ (or some other given set of ratios).

And so on, ...; the variety is unlimited. One naturally asks: *For which kinds of problems will such an approach work?*

The reader is urged to tackle these five problems, and to formulate others of their kind. They are sources of rich investigations

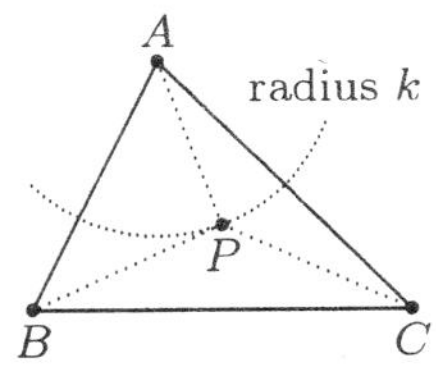

Figure 1.4. *Which point P minimizes $PA + PB + PC$?*

and will repay, in full measure and greater, the time and resources spent in uncovering their secrets.

★ ★ ★

It is useful at this point to state a general principle.

> When a particular strategy does work, (a) try to find out what makes it work (it may be some special feature or peculiarity that distinguishes the problem); and (b) try to list other problems for which the same strategy might work.

With regard to (b), it seems unclear whether the approach described for the '$OA = OB = OC$' problem would work for the following problem:

> **Problem**
> *Given three points A, B, C, find the point(s) P for which the sum $PA + PB + PC$ has the lowest possible value.*

(But consider this: suppose that the distance PA is kept fixed, say $PA = k$ for some k, so that P lies on a circle with center A and radius k. For which point P on this circle is $PB + PC$ minimized? (See Figure 1.4.)

It is not hard to guess that the best we can do is to choose P such that $\angle APB = \angle APC$. Now, we relax the restriction that $PA = k$. What conclusion follows from this? Will this approach yield a solution to the problem?)

1.8 Triangle problem

> **Problem**
> *In $\triangle ABC$, show that if angle A is twice angle B, then $a^2 = b(b + c)$. (Here, a, b, c are the lengths of the sides BC, CA, AB; see Figure 1.5(a).)*

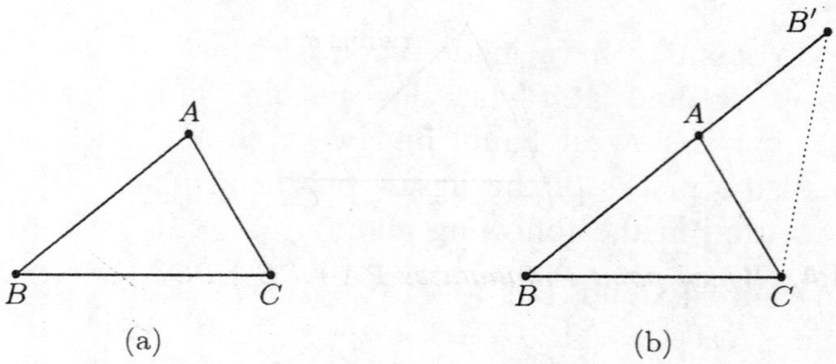

Figure 1.5. *If* $\angle A = 2\angle B$, *then* $a^2 = b(b+c)$.

COMMENT. 'Dimensionally' the condition $a^2 = b(b+c)$ makes sense—its terms all have the same degree, namely 2. That is, the condition is a *homogeneous* one.

How do we proceed? Consider the condition $a^2 = b(b+c)$. Can we make it look more familiar? We start by writing it thus: $a/b = (b+c)/a$. This suggests an approach using similar triangles, but where are the similar triangles, and where is the length $b+c$?

There are two natural ways in which the length $b+c$ can occur in the diagram, namely:

1. by extending BA to B' so that $AB' = AC$, in which case $BB' = BA + AB' = c + b$ (see Figure 1.5(b));
2. by extending CA to C' so that $AC' = AB$, in which case $CC' = b + c$ (please draw the figure yourself).

We shall explore both possibilities.

- *Possibility 1.* In Figure 1.5(a) the condition to be proved reads:

$$\frac{BC}{AC} = \frac{BB'}{BC}.$$

This looks like an equality of ratios arising from a pair of similar triangles—but not quite; the angles do not correspond in the correct manner. If we rewrite the equation as $BC/BB' = AC/BC$, the angles still do not correspond.

But now note that $B'C$ and BC are equal: $\angle BAC = \angle AB'C + \angle ACB' = 2\angle AB'C$, $\therefore\ \angle AB'C = \frac{1}{2}\angle BAC = \angle ABC$, $\therefore\ \angle B'BC = \angle BB'C$, $\therefore\ BC = B'C$. With one BC replaced by $B'C$, the condition to be proved now becomes $BC/AC = BB'/B'C$, or

$$\frac{BC}{BB'} = \frac{AC}{B'C},$$

and we do seem to have made some progress—the similarity of triangles $B'BC$ and ACB' has now become more plausible. The final solution is now at hand, and we complete the solution by putting all the pieces of the jigsaw puzzle in place. The solution may be written in the following manner.

Construction Extend BA to B' so that $AB' = AC$; join $B'C$. Note that $\angle BB'C = \angle ACB' = A/2$.

Proof Using the condition $\angle A = 2\angle B$ and the equality $\angle BAC = \angle AB'C + \angle ACB'$, we obtain $\angle CBB' = \angle CB'B = \angle ACB'$ and $CB = CB' = a$, so $\triangle CAB' \sim \triangle BCB'$ (both triangles are isosceles, and with the same base angles). It follows that

$$\frac{CB'}{CA} = \frac{BB'}{BC}.$$

In other words, $a/b = (b+c)/a$, or $a^2 = b(b+c)$. □

Here is a trigonometric version of this proof. Let $\angle B = \theta$; then $\angle A = 2\theta$ and $\angle B' = \theta$. The sine rule ("the sides of a triangle are proportional to the sines of the angles opposite them") as applied to $\triangle ABC$ and $\triangle B'BC$ yields

$$\frac{a}{b} = \frac{\sin 2\theta}{\sin\theta}, \qquad \frac{b+c}{a} = \frac{\sin 2\theta}{\sin\theta}.$$

Therefore, $a/b = (b+c)/a$, and the result is immediate. □

A trigonometric approach that avoids the construction of auxiliary lines is the following. Since $\angle B = \theta$ and $\angle A = 2\theta$, we get $\angle C = 180° - 3\theta$, so $\sin C = \sin(180° - 3\theta) = \sin 3\theta$, and

$$a : b : c = \sin 2\theta : \sin\theta : \sin 3\theta.$$

Using the two identities $\sin 2\theta = 2\sin\theta\cos\theta$ and $\sin 3\theta = 3\sin\theta - 4\sin^3\theta$, we get $a : b : c = 2\cos\theta : 1 : 3 - 4\sin^2\theta$, and so $a : b : b + c = 2\cos\theta : 1 : 4(1 - \sin^2\theta)$, or

$$a : b : b + c = 2\cos\theta : 1 : 4\cos^2\theta.$$

This immediately yields $a^2 = b(b+c)$. □

The reader may complain – with reason! – that the manner in which the solution has been given is rather confusing. But surely this reflects our actual experience in problem-solving? Typically, we do not proceed in a clear-cut, precise and logical manner; things do not happen in an inevitable and sure fashion. In actuality, the approach route may be confused, imprecise and convoluted, and it blunders along, avoiding a pitfall here, seeing a worthwhile fact there, recording it, pursuing it, and so on. After proceeding for a while, with some luck, one glimpses the final 'big picture'; this is the moment of insight, which Martin Gardner calls the 'AHA!' moment, and if it comes through, the rest is easy. (Well, not always; even if the final picture has been glimpsed, the details – the "nitty-gritty" – have still to be worked out. Often, this is the difficult part.) This is the actual manner in which research in mathematics takes place—let no one fool you on this matter! But many researchers have a passion (as foxes do) for erasing their tracks, the tracks in this case being a record of the initial blundering around. So, they work hard at presenting a version of their discoveries that omits all reference to their initial false starts. The final version is like a jewel, complete and dazzling in itself, but with absolutely no clue as to how anyone could have created it.

- *Possibility 2.* We now consider possibility 2, where the diagram is different; the length $b+c$ appears in another place; see Figure 1.5(b). The approach appears unclear now, with no similar triangles in sight; but a solution is at hand here too! We recall the following theorems about circles (please prove them yourself):

Theorem 1 *If PT is a tangent to a circle Γ at T, and TA is a chord of Γ, then the angle between tangent and chord, namely $\angle PTA$, equals the angle subtended by chord TA in the "alternate segment". That is, if B is a point on the circle such that P and B are on opposite sides of line AT, then $\angle PTA = \angle TBA$.* (See Figure 1.6.)

Theorem 2 *Given a circle Γ and a triangle ABT inscribed in it, if PT is a line with P and B on opposite sides of line AT such that* angle $PTA =$ angle ABT, *then PT is tangent to the circle.* (See Figure 1.6; this is the converse of Theorem 1.)

Theorem 3 *Given a point P point outside a circle Γ, a line through P intersecting the circle at A and B, and a line through P tangent to the circle at T, the following relation holds between the lengths PT, PA and PB:* (See Figure 1.7:)

$$PT^2 = PA \cdot PB.$$

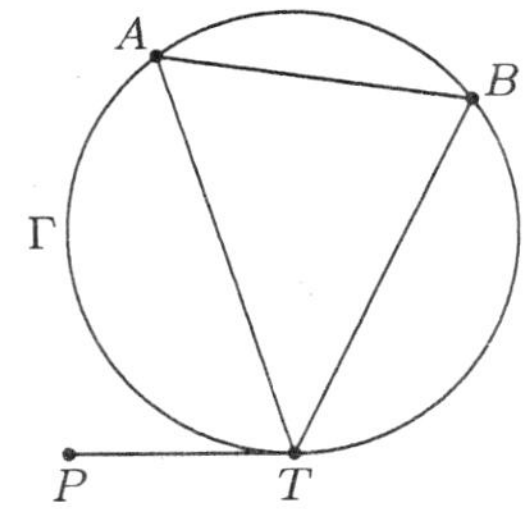

Figure 1.6. *The alternate angle theorem:* angle PTA = angle TBA

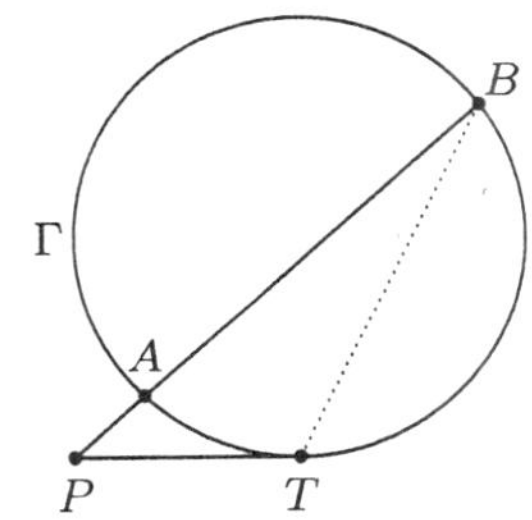

Figure 1.7. *The chord theorem:* $PT^2 = PA \times PB$

In Figure 1.8, angle BAC is twice angle ABC, and side CA is extended to C' such that $AC' = AB$. Denote angle ABC by θ, then $\angle BAC = 2\theta$. Since $\angle BAC = \angle C'BA + \angle BC'A$, and $\angle C'BA = \angle BC'A$ (by virtue of the equality $AB = AC'$), it follows that $\angle C'BA = \angle BC'A = \theta$ and therefore that $\angle ABC = \angle BC'A$. Let Γ be the circumcircle of $\triangle C'BA$. The angle equality just stated implies that CB is tangent to the circle (Theorem 2, above). Now we invoke Theorem 3 to conclude that:

$$CB^2 = CA \cdot CC', \qquad \therefore \quad a^2 = b(b+c).$$

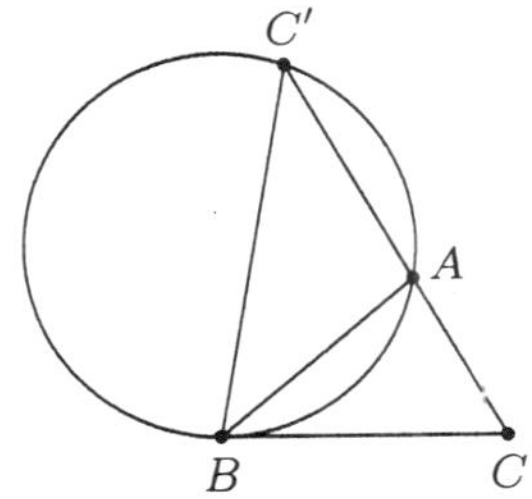

Figure 1.8. *Solution of the* $a^2 = b(b+c)$ *problem*

And the deed is done! □

Remarkably, our story has a postscript—the converse statement is also true. See if you can prove it!

Theorem *If $\triangle ABC$ has $a^2 = b(b+c)$, then angle A is twice angle B.*

★ ★ ★

Analogous to the relation $a^2 = b(b+c)$, which holds for triangle ABC when angle A is twice angle B, there is a relation which holds between the sides a, b, c when angle A is *three* times angle B; only, it is more complicated. The relation is:

$$b\,c^2 = (a+b)\,(a-b)^2.$$

Observe that the dimensionality check works out: the condition is homogeneous, of degree 3. At the first glance, it looks as though the 'pure geometry' approach is doomed; there seems no natural way whereby third degree terms can be made to appear through a purely geometric analysis. So we use trigonometry, and this makes the problem much easier.

Let θ be the measure of $\angle B$ in degrees, then $\angle B = 3\theta$ and $\angle C = 180 - 4\theta$. Using the sine rule and the relations $\sin 3\theta = 3\sin\theta - 4\sin^3\theta$ and $\sin 4\theta = 4\sin\theta\cos\theta(2\cos^2\theta - 1)$, we have

$$\begin{aligned} a : b : c &= \sin 3\theta : \sin\theta : \sin(180 - 4\theta) \\ &= \sin 3\theta : \sin\theta : \sin 4\theta \\ &= 3\sin\theta - 4\sin^3\theta : \sin\theta : 4\sin\theta\cos\theta(2\cos^2\theta - 1) \\ &= 3 - 4\sin^2\theta : 1 : 4\cos\theta(2\cos^2\theta - 1) \\ &= 4\cos^2\theta - 1 : 1 : 4\cos\theta(2\cos^2\theta - 1). \end{aligned}$$

It follows from this that $(a+b) : b : c = 4\cos^2\theta : 1 : 4\cos\theta(2\cos^2\theta - 1)$ and, therefore,

$$\begin{aligned} a + b &= 4\cos^2\theta\, b, \qquad \therefore\ 2\cos\theta = \sqrt{\frac{a+b}{b}}, \\ c &= 4\cos\theta(2\cos^2\theta - 1)b = 2\sqrt{\frac{a+b}{b}}\left(\frac{a+b}{2b} - 1\right)b \\ &= \sqrt{\frac{a+b}{b}}(a-b), \qquad \therefore\ c^2 = \frac{a+b}{b}(a-b)^2. \end{aligned}$$

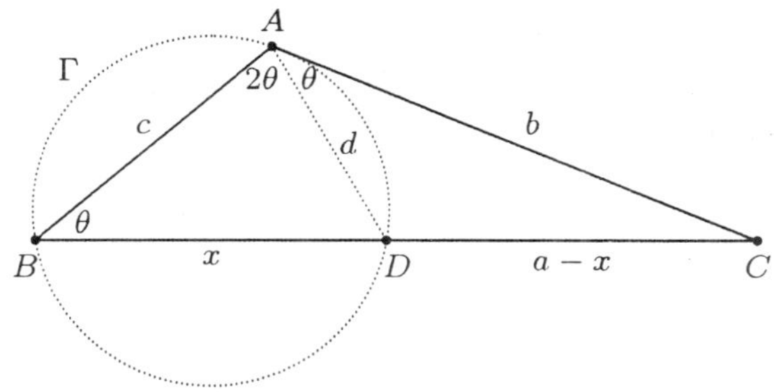

Figure 1.9. *Angle A is 3 times angle B.*

From the last equation we obtain $bc^2 = (a+b)(a-b)^2$. □

So, is the trigonometric approach the only one possible for this problem? Not quite!—one can find a geometric proof yet again, making use of the earlier results. In triangle ABC, (see Figure 1.9) we have $\angle B = \theta$ and $\angle A = 3\theta$. We start by locating a point D on BC such that $\angle CAD = \theta$. Then $\angle BAD = 2\theta$, so in triangle BAD we have $\angle BAD = 2\angle ABD$. Also, since $\angle CAD = \angle ABD$, the circumcircle Γ of triangle ABD is tangent to AC at A (Theorem 3, above); see Figure 1.9. Let d and x be the lengths of AD and BD, respectively; then, $DC = a - x$. We reason as follows below.

Since CA is tangent to Γ at A, $CA^2 = CD \cdot CB$. In $\triangle ABD$, since $\angle BAD = 2\angle ABD$, $BD^2 = AD \cdot (AD + AB)$. Finally, since $\angle ADC = 3\theta$, $\triangle ABC \sim \triangle DAC$. So, we have the following three relations:

$$\begin{aligned} b^2 &= a(a-x), \\ x^2 &= d(d+c), \\ a : b : c &= b : a - x : d. \end{aligned}$$

The first relation yields $x = a - b^2/a = (a^2 - b^2)/a$, and from the third relation we obtain $d = bc/a$. Substituting for x and d in the second relation, we obtain

$$\frac{(a^2-b^2)^2}{a^2} = \frac{bc}{a}\left(\frac{bc}{a} + c\right).$$

The denominators (a^2) are the same on both sides, and $a+b$ is a common factor. Cancelling off the common terms, which are clearly non-zero, we are left with the relation

$$(a-b)^2(a+b) = bc^2,$$

which was to be proved. □

The moral of the story is: *Don't give up hope; and don't get stuck to any particular approach!*

★ ★ ★

It is not hard to see how a progression of such results may be found. In the general case, we suppose that if in $\triangle ABC$ we have $\angle A = n\angle B$, where n is a positive integer, then we have some relation between a, b and c, say

$$f_n(a,b,c) = 0.$$

The function f_n may be taken to be homogeneous. For example, when $n=2$ we have $a^2 = b(b+c)$, so we may write $f_2(a,b,c) = a^2 - b(b+c)$.

We establish a connection between f_n and f_{n+1} by drawing a figure similar to Figure 1.9, but with 2θ replaced by $n\theta$. In $\triangle ABC$, we have $\angle BAC = (n+1)\angle ABC = (n+1)\theta$. Let D be a point on BC such that $\angle DAC = \theta$. Then $\angle BAD = n\theta$ and $\angle ADC = (n+1)\theta$. Let $AD = d$, $BD = x$; then $DC = a - x$; $\triangle ABC \sim \triangle DAC$; and CA is tangent to the circumcircle Γ of $\triangle ABD$. Therefore,

$$b^2 = a(a-x), \quad \therefore \; x = a - b^2/a = (a^2-b^2)/a,$$
$$a : b : c = b : a - x : d, \quad \therefore \; d = bc/a.$$

In $\triangle ABD$, since $\angle BAD = n\angle ABD$, the condition f_n applies to the lengths of its sides; that is, $f_n(x,d,c) = 0$. Substituting for x and d, we get

$$f_n\left(\frac{a^2-b^2}{a}, \frac{bc}{a}, c\right) = 0.$$

By invoking the homogeneity of f_n we may replace the above condition by: $f_n\left(a^2-b^2, bc, ca\right) = 0$. This condition, when simplified, yields the 'next' condition: $f_{n+1}(a,b,c) = 0$.

When $n=1$ we have $\angle A = \angle B$, $\therefore a = b$; so we take $f_1(a,b,c) = a - b$. The condition for $n=2$ is then:

$$f_1\left(a^2-b^2, bc, ca\right) = 0, \text{ or } a^2 - b^2 - bc = 0.$$

This simplifies to $a^2 = b(b+c)$, as expected. So, we take f_2 to be given by $f_2(a,b,c) = a^2 - b(b+c)$. To move to the case $n = 3$, we write

$$f_2\left(a^2 - b^2, bc, ca\right) = 0, \text{ or } \left(a^2 - b^2\right)^2 - bc(bc + ca) = 0.$$

This simplifies to $(a-b)^2(a+b) - bc^2 = 0$, and we take f_3 to be given by

$$f_3(a,b,c) = (a-b)^2(a+b) - bc^2.$$

For $n = 4$, we write $f_3\left(a^2 - b^2, bc, ca\right) = 0$, or

$$\left(a^2 - b^2 - bc\right)^2\left(a^2 - b^2 + bc\right) - a^2bc^3 = 0.$$

After some hard work, this simplifies to the fourth degree condition $(a-b)(a+b)(a^4 - 2a^2b^2 - a^2bc + b^4 + b^3c - b^2c^2 - bc^3) = 0$. Since $a+b$ and $a-b$ are clearly non-zero (recall that angle $A = 4$ angle B), we may drop the first two factors; we are left with:

$$a^4 - 2a^2b^2 - a^2bc + b^4 + b^3c - b^2c^2 - bc^3 = 0.$$

This progression of results may be continued indefinitely. But now it is time for us to move on ....

1.9 Prison blues

Our next example is a well-known one, so much so that one feels rather embarrassed to quote it here. Nevertheless, we shall do so, if only because the problem and solution are so pretty.

Problem
In a certain prison, a large number of prisoners are housed in a long corridor of cells. Obeying the terms of an amnesty, the jailor releases certain prisoners by performing the following actions. He first walks down the corridor and unlocks every door. (The inmates are all asleep while he does this; no one escapes!) Next, he returns to his starting point and walks down the corridor again; this time he locks every *second* door. He then returns once again to the starting point and walks down the corridor yet again; this time he turns the key of every *third* door (unlocking it if it is locked, locking it if it is unlocked). He continues this set of actions, stopping next at every fourth door, then every fifth door, every sixth door, ...; each time he simply turns the key in the door. Ultimately, certain cells are left open by this procedure; *which ones*?

Observe that the actions performed on trip $\#d$ affect only the cells numbered d, $2d$, $3d$, $4d$, ... ; in short, the cells numbered by a multiple of d. Now, the cells are alternately unlocked and locked; each turn of the key reverses its status. At the start each cell is locked, so for a cell to be ultimately left open, it must undergo an *odd* number of key-reversals. Or, since each reversal corresponds to a divisor of the cell number, *the cell number must have an odd number of divisors.* Our task thus reduces to identifying the numbers that have an odd number of divisors.

For an arbitrary positive integer n, let $d(n)$ denote its number of divisors (including 1 and n itself); thus, $d(10) = 4$, because the divisors of 10 are 1, 2, 5 and 10, and $d(20) = 6$, because the divisors of 20 are 1, 2, 4, 5, 10 and 20. Here are a few sample values of $d(n)$.

n	1	2	3	4	5	6	7	8	9	10
$d(n)$	1	2	2	3	2	4	2	4	3	4

For which n is $d(n)$ odd? We see that for n lying between 1 and 10, $d(n)$ is odd only for $n = 1$, 4 and 9. Let us now see what happens for $11 \le n \le 20$.

n	11	12	13	14	15	16	17	18	19	20
$d(n)$	2	6	2	4	4	5	2	6	2	6

We see that in this case, $d(n)$ is odd only for $n = 16$. So, we have a set of 'successful' numbers which starts $\{1, 4, 9, 16, \ldots\}$. The pattern is unmistakable: *$d(n)$ appears to be odd precisely when n is square.* Is this true in general? We shall do some computations for a few more values of n, say for the numbers in the range $61 \le n \le 70$.

n	61	62	63	64	65	66	67	68	69	70
$d(n)$	2	4	6	7	4	8	2	6	4	8

(Check that the figures are correct; e.g., $d(70) = 8$, because 70 has the following eight divisors: 1, 2, 5, 7, 10, 14, 35 and 70.)

We see that for n lying between 61 and 70, $d(n)$ is odd only for $n = 64$; once again, this is the only square number in the range 61, ..., 70. Our hypothesis does seem to be correct.

★ ★ ★

It remains now to *prove* that the hypothesis is correct, and this is accomplished as follows. Let n be a positive integer and let d be a divisor of n. Since $d \cdot n/d = n$, it follows that n/d too is a divisor of n. Thus, each divisor d of n has a mate, namely n/d, which is distinct from d provided that $d \neq n/d$, i.e., provided that $d^2 \neq n$ or $d \neq \sqrt{n}$. (The mate of n/d is d itself.) We infer from this that if n is not square, then $d(n)$ is *even*, because the divisors get paired off as described (that is, d gets paired with n/d).

EXAMPLE. For $n = 80$, the divisors get paired thus: $\{1, 80\}$, $\{2, 40\}$, $\{4, 20\}$, $\{5, 16\}$, $\{8, 10\}$; therefore, $d(80) = 2 \times 5 = 10$, an even number.

On the other hand, if n is square, then $d(n)$ is necessarily odd, because each divisor $d \neq \sqrt{n}$ has a mate n/d different from itself, while the divisor $\sqrt{n}$ stands alone, with no mate. The presence of the 'odd man out' makes $d(n)$ an odd number.

We have thus proved our guess: *$d(n)$ is odd if and only if n is square.* Thus, the lucky prisoners are those whose cell numbers are squares. Note that *en route* to the solution, we have proved a result in number theory.

An alternative proof There is another way of showing that $d(n)$ is odd if and only if n is square. Let the prime factorization of n be

$$n = p^a\, q^b\, r^c \cdots,$$

where $p, q, r, \ldots$ are primes and $a, b, c, \ldots$ are positive integers. Then, each divisor d of n has the form

$$d = p^{a'}\, q^{b'}\, r^{c'} \cdots,$$

where $a', b', c' \ldots$ are integers with $0 \leq a' \leq a$, $0 \leq b' \leq b$, $0 \leq c' \leq c$, ... This is so because a divisor of n is necessarily composed of the same primes as n itself, and no prime can occur to a higher power in the divisor than in n. By letting $a', b', c', \ldots$ take all possible values between their respective limits, we obtain all possible divisors of n. Now, observe that a' can take $a+1$ different values, namely 0, 1, 2, ..., a; likewise, b' can take $b+1$ different values, c' can take $c+1$ different values, and so on. It follows that the number of different values that d can take is

$$(a+1)(b+1)(c+1)\cdots$$

Therefore, $d(n) = (a+1)(b+1)(c+1)\cdots$. It is easy to verify the correctness of this formula.

EXAMPLE. For $n = 80 = 2^4 \cdot 5$, the formula gives $d(10) = (4+1)\cdot(1+1) = 5\cdot 2 = 10$, and for $n = 30 = 2\cdot 3\cdot 5$, it gives $d(30) = 2\cdot 2\cdot 2 = 8$. Both statements are correct.

Now, the product $(a+1)(b+1)(c+1)\cdots$ is odd if and only if *each* of the numbers $(a+1)$, $(b+1)$, $(c+1)$, ... is odd; in other words, if and only if each of the numbers $a, b, c, \ldots$ is even. But if $a, b, c, \ldots$ are even, then n is clearly square; indeed,

$$n = \left(p^{a/2}\, q^{b/2}\, r^{c/2} \cdots\right)^2.$$

We conclude, as earlier, that $d(n)$ is odd if and only if n is square. □

COMMENT. While this proof is longer and decidedly less pretty than the first one, it also yields more ("More is less", as Polya was fond of saying). Thus, it enables us to classify the values of n for which $d(n) = 10$ (these are the numbers of the form $p^4 q$ or p^9); or the values of n for which $d(n)$ is a power of 2 (these are the numbers n for which the exponent of each prime factor of n is 1 less than a power of 2, e.g., $n = pq^7$ or $n = p^3 q^{15}$ for distinct primes p and q); and so on.

1.10 Notes from Polya

The moral of these anecdotes is that we must not fight shy of plunging into a problem with a broadsword in hand—be it in formulating a hypothesis, or in pushing a hypothesis to its logical limits—to a point where its true worth can be judged. One must not be overly concerned with niceties in the initial stages of the battle. This holds true not only in mathematics but also in science, chess and – though this may sound unlikely – poetry. The final polished version of a theorem or piece of poetry that you see in print is the result of months (perhaps years!) of labour after the initial insight has been gained. It is *always* the case that insight precedes investigation.

In the pages to come, you will encounter many problems. Of these, some have a recreational or non-serious touch to them. Rest assured that they are not mere diversions! The approach at bottom is the same, whether for serious mathematics or for 'non-serious' mathematics.[4] You will also encounter problem sequences, in which successively harder versions of a problem are presented; sometimes they may be generalizations of a theme ("variations on a theme", as they say in music). Here the object is to demonstrate the process of generalization, a theme that characterizes so much of modern mathematics.

The student may ask at this point: *For the professional mathematician, where do the problems come from in the first place?* This question can also be asked of those who create problems for competitions in mathematics—say, the Olympiads. There is no easy answer to the question. The nature of research is such that while investigating one problem, others arise in a natural manner; mathematicians often talk of problems that 'pose themselves'. In such cases, it may happen that the mathematician temporarily shelves the original problem and instead pursues another one, a 'side-branch', as it were. Research is anything but linear, with a fixed, directed goal in mind. Anyone who diligently pursues problems is bound to reach this conclusion.

⋆ ⋆ ⋆

We shall end this section with some excerpts from one of the most outstanding contributions to the study of problem-solving: George Polya's *How to Solve It*.

"... A great discovery solves a great problem but there is a grain of discovery in the solution of any problem. Your problem may be modest; but if it challenges your curiosity and brings into play your inventive faculties, and if you solve it by your own means, you may experience the tension and enjoy the triumph of discovery ...." [from the Preface]

"... Solving problems is a practical skill like, let us say, swimming. We acquire any practical skill by imitation and practice. Trying to swim, you imitate what other people do with their hands and feet to keep their heads above water, and,

[4]Indeed, is not the term *Recreational Mathematics* superfluous? Is not all mathematics recreational? Well, the answer is both yes and no. In terms of the motivations of individual mathematicians, it may well be true that at bottom all mathematics is recreational. But the remarkable thing about mathematics is that it has a way, all of a sudden, of becoming useful. Physicists, engineers, chemists, economists, linguists, demographers, and so on, end up finding uses for concepts devised by mathematicians. So, at some stage, the subject suddenly becomes serious! There is a touch of mystery about how or why this happens, and no one has found a satisfactory way of explaining it.

finally, you learn to swim by practicing swimming. Trying to solve problems, you have to observe and imitate what other people do when solving problems and finally, you learn to do problems by doing them ...." [page 4]

"... It is hard to have a good idea if we have little knowledge of the subject, and impossible to have it if we have no knowledge. Good ideas are based on past experience and formerly acquired knowledge. Mere remembering is not enough for a good idea ..." [Polya points out here that a good base of knowledge of problems and approaches that have succeeded or failed in the past is often vital.] "... Thus, it is often appropriate to start the work with the question: Do you know a related problem? ...." [page 9]

"... Analogy pervades all our thinking, our everyday speech and our trivial conclusions as well as artistic ways of expression and the highest scientific achievements. Analogy is used on very different levels. People often use vague, ambiguous, incomplete or incompletely clarified analogies, but analogy may reach the level of mathematical precision. All sorts of analogy may play a role in the discovery of the solution and so we should not neglect any sort .... We may consider ourselves lucky when, trying to solve a problem, we succeed in discovering a simpler analogous problem ...." [pages 37,38]

"... If you cannot solve the proposed problem do not let this failure afflict you too much but try to find consolation with some easier success, try to solve first some related problem; then you may find courage to attack your original problem again. Do not forget that human superiority consists in going around an obstacle that cannot be overcome directly, in devising some suitable auxiliary problem when the original one appears insoluble ...." [page 114]

"... In mathematics as in the physical sciences we may use observation and induction to discover general laws. But there is a difference. In the physical sciences, there is no higher authority than observation and induction but in mathematics there is such an authority: rigorous proof.... After having worked a while experimentally ... and discovered an interesting result ... [we need to] ... try to establish it definitively by a rigorous proof." [page 117]

"... The future mathematician should be a clever problem-solver; but to be a clever problem-solver is not enough. In due time, he should solve significant mathematical problems; and first he should find out for which kind of problems his native gift is particularly suited .... For him, the most important part of the work is to look at the completed solution. Surveying the course of his work and the final shape of the solution, he may find an unending variety of things to observe. He may meditate upon the difficulty of the problem and about the decisive idea, he may try to see what hampered him and what helped him ....

He may try to clarify his present problem by comparing it to problems formerly solved; he may try to invent new problems which he can solve on the basis of his just completed work .... Digesting the problems he solved as completely as he can, he may acquire well ordered knowledge, ready for use ...." [page 205]

"... The future mathematician learns, as does everybody else, by imitation and practice. He should look out for the right model to imitate. He should observe a stimulating teacher ... he should read not only current textbooks but good authors till he finds one whose ways he is naturally inclined to imitate. He should enjoy and seek what seems to him simple or instructive or beautiful. He should solve problems, choose the problems which are in his line, meditate upon their solution, and invent new problems. By these means, and by all other means, he should endeavour to make his first important discovery: he should discover his likes and his dislikes, his taste, his own line ...." [pages 205-206]

"... The intelligent problem-solver tries first of all to understand the problem as fully and as clearly as he can. Yet understanding alone is not enough; he must concentrate upon the problem, he must desire earnestly to obtain its solution. If he cannot summon up real desire for solving the problem he would do better to leave it alone. The open secret of real success is to throw your whole personality into your problem ...." [page 207]

⋆ ⋆ ⋆

Probably there is no better note to end this chapter than with these words. The overture is over, and we now present to the reader the main body of the work—the problems themselves.

Chapter 2

Number Play

2.1 Surds

This section could alternatively be titled: *Playing with a Four-Function Calculator*, the 'four functions' being the rational operations of arithmetic: $+$, $-$, $\times$, $\div$. The keys available on the calculator are those corresponding to these operations, and (naturally) those for '=', '`Clear`', '`All Clear`', memory storage, memory recall. The square-root and power function keys are not available, nor those associated with the trigonometric, logarithmic and exponential functions. (It goes without saying that we do not have access either to a slide rule or to a table of logarithms!)

The question we now pose is this: *Using only these four operations, how do we compare numbers whose definitions involve the disallowed operations?* That is, how do we compare irrational quantities if only rational arithmetic operations are permitted? The problems listed below will make the scope of the question clear.

EXERCISES

2.1.1 For each pair of numbers given below, determine which one is larger.

(a) $\sqrt{3}+\sqrt{7}, \quad 2\sqrt{5}$;
(b) $\sqrt{11}+\sqrt{19}, \quad 2\sqrt{15}$;
(c) $\sqrt{21}+\sqrt{29}, \quad \sqrt{23}+\sqrt{27}$;
(d) $\sqrt{31}+\sqrt{34}, \quad \sqrt{32}+\sqrt{33}$;

(e) $\sqrt{a}+\sqrt{a+3}$, $\quad \sqrt{a+1}+\sqrt{a+2}$, where $a > 0$.

Frame a suitable hypothesis on the basis of your findings.

2.1.2 For each pair of numbers given below, determine which one is larger.

(a) $2^{1/2}$, $\quad 3^{1/3}$;
(b) $3^{1/3}$, $\quad 4^{1/4}$;
(c) $4^{1/4}$, $\quad 5^{1/5}$;
(d) $5^{1/5}$, $\quad 6^{1/6}$;
(e) $6^{1/6}$, $\quad 7^{1/7}$.

Frame a suitable hypothesis on the basis of your findings.

2.1.3 Our objective in the following sequence of problems will be to determine good approximations to square roots. First, we shall explain what we mean by "good approximation".

EXAMPLE. Since $\sqrt{2} \approx 1.4142$ and $7/5 = 1.4$, we may feel that $7/5$ is a good approximation to $\sqrt{2}$. However, 'good' is a rather imprecise and ambiguous word, so we opt for a comparative terminology: *better approximation*.

EXAMPLE. Consider the approximations $3/2$ and $7/5$ to $\sqrt{2}$. Of these, $7/5$ is the better one, because

$$\left|\sqrt{2}-\frac{7}{5}\right| \approx 0.0142, \qquad \left|\sqrt{2}-\frac{3}{2}\right| \approx 0.0857.$$

So, $|\sqrt{2}-3/2|$ is nearly 6 times as large as $|\sqrt{2}-7/5|$.

Problem Show that $7/5$ is a better approximation to $\sqrt{2}$ than $3/2$, *without* using the information that $\sqrt{2} \approx 1.4142$. Remember that you only possess a 4-function calculator.

2.1.4 Use the fact that $|\sqrt{2}-7/5| \approx 6 \times |\sqrt{2}-3/2|$ to arrive at an approximation to $\sqrt{2}$ which is superior to $7/5$.

2.1.5 Consider the following assertion:

$$\text{If } 0 < x < 1, \text{ then } 0 < x^2 < x.$$

To put it in words: *If x lies between 0 and 1, then x^2 lies* *between 0 and x.* This implies the following (obtained by repeated squaring):

$$\text{If } 0 < x < 1, \text{ then } 1 > x > x^2 > x^4 > x^8 > \cdots \geq 0.$$

Thus, x^{2^n} decreases steadily as n increases. Indeed, $x^{2^n} \to 0$ as $n \to \infty$.

Apply this principle to the number $\sqrt{2} - 7/5$ to find good approximations to $\sqrt{2}$. (Use the fact that $\sqrt{2} - 7/5$ lies between 0 and 1.) Amongst the fractions whose denominator is less than 100, which one is closest to $\sqrt{2}$?

2.1.6 Find good rational approximations to $\sqrt{3}$, $\sqrt{5}$ and $\sqrt{10}$. (A *rational number* is a fraction, that is, a number of the form a/b where a, b are coprime integers with $b > 0$.)

Amongst the fractions whose denominator is less than 100, identify the ones closest to $\sqrt{3}$, $\sqrt{5}$ and $\sqrt{10}$.

2.1.7 Prove that if $\dfrac{m}{n}$ is close to $\sqrt{2}$, then $\dfrac{m}{2n} + \dfrac{n}{m}$ is still closer.

EXAMPLE. Starting with the fraction 7/5, which is close to $\sqrt{2}$, we obtain the fraction

$$\frac{7}{10} + \frac{5}{7} = \frac{99}{70},$$

which is much closer. Repeating the procedure with 99/70, we obtain

$$\frac{99}{140} + \frac{70}{99} = \frac{19601}{13860}.$$

The fraction on the right side differs from $\sqrt{2}$ by less than 2×10^{-9}! (For comparison: $\sqrt{2} = 1.414213562\ldots$ and $19601/13860 = 1.414213564\ldots$.)

2.1.8 Formulate and prove statements analogous to the one made above, but with $\sqrt{3}$ and $\sqrt{5}$ in place of $\sqrt{2}$. (*"If m/n is close to $\sqrt{3}$, then ... is still closer."*)

2.1.9 Show that for any positive integer n, the quantities $\sqrt{n+1}$ and $\sqrt{n}$ differ by less than $1/2\sqrt{n}$.

EXAMPLE. $\sqrt{101} - \sqrt{100} \approx 0.0499$, which is less than $1/20$. (This corresponds to $n = 10$.)

2.1.10 Let a_n denote the quantity $n+\frac{1}{2}-\sqrt{n^2+n}$. Sample calculations suggest that a_n diminishes steadily to 0 as n increases. Let $b_n = na_n$. The following table gives a few sample values of b_n.

n	10	100	1000	10000
b_n	0.11912	0.12438	0.12494	0.12499

Observe that $b_n \approx 1/8$ for large n. Why should this be so?

2.1.11 Show that $7^{\sqrt{5}} > 5^{\sqrt{7}}$. (No calculators, please!)

2.1.12 Is it true that $(n+2)^{\sqrt{n}} > n^{\sqrt{n+2}}$ for all positive integers n?

2.1.13 Which is larger, $\sqrt[3]{60}$ or $2+\sqrt[3]{7}$?

2.1.14 Find the least square whose base-10 representation begins with the digits 1234567.

2.1.15 Find the least cube whose base-10 representation begins with the digits 1234567.

2.2 Excellent approximations

If we push the idea of a "good approximation" a little further, we discover something quite fascinating.

Let $x > 0$ be irrational. We are interested in finding *excellent rational approximations* to x, 'excellent' being defined as below.

> **"Excellent approximation"** The fraction a/b is said to be an "excellent approximation" to the number x if amongst all the fractions p/q where p, q are integers with $0 < q \leq b$, the one for which $|p - qx|$ is the *least* is a/b. (Here, $|x|$ refers to the absolute value of x; for example, $|-2.1| = 2.1$.)

EXAMPLE. The fraction 3/2 is an excellent approximation for $\sqrt{2}$, because $|3-2\sqrt{2}|$ is less than each of the numbers $|2-2\sqrt{2}|$, $|1-2\sqrt{2}|$, $|2-\sqrt{2}|$, ... However, 4/3 is not excellent, because $|4-3\sqrt{2}| > |3-2\sqrt{2}|$ (this is equivalent, after squaring, to $17 > 12\sqrt{2}$, which is true since $289 > 288$); so 3/2 has 'beaten' 4/3. However, 7/5 is an excellent approximation, as can be checked by comparing 7/5 with 7/4, 3/2, and so on.

It is not hard to anticipate that infinitely many excellent approximations exist to any given irrational number.

REMARK 1. As p and q take different integer values, the values taken by $|p - qx|$ are distinct; no repetitions occur. (The repetition of a value would imply that x is rational; for if $p - qx = \pm(p' - q'x)$, then by solving the equation for x we would get a rational value for x.) Thus, the minimum referred to in the definition occurs uniquely, and the term 'excellent approximation' is well-defined.

REMARK 2. In view of Remark 1, corresponding to any fixed irrational number x, we may define a 'total order' on the set of all fractions. That is, given any two distinct fractions a/b and c/d, we say that $a/b \prec c/d$ if $|a - bx| < |c - dx|$, or else we say that $c/d \prec a/b$. (We may also write $a/b \succ c/d$ in this case.) The relation '$\prec$' imposes what is termed as a *total order* on the rational numbers.

To remind ourselves that the ordering is with reference to some particular value of x, the value may be written on top of the '$\prec$' symbol, thus: $\overset{\sqrt{2}}{\prec}$. For example, we have

$$\frac{1}{1} \overset{\sqrt{2}}{\prec} \frac{2}{1}.$$

EXAMPLE. With $x = \sqrt{2}$, the fractions 1/1, 2/1, 3/2, 4/3, 5/4, 6/5, 7/5, 8/5, 9/7 and 10/7 may be arranged in the following order:

$$\frac{7}{5} \overset{\sqrt{2}}{\prec} \frac{10}{7} \overset{\sqrt{2}}{\prec} \frac{3}{2} \overset{\sqrt{2}}{\prec} \frac{4}{3} \overset{\sqrt{2}}{\prec} \frac{1}{1} \overset{\sqrt{2}}{\prec} \frac{2}{1} \overset{\sqrt{2}}{\prec} \frac{5}{4} \overset{\sqrt{2}}{\prec} \frac{9}{7} \overset{\sqrt{2}}{\prec} \frac{8}{5} \overset{\sqrt{2}}{\prec} \frac{6}{5}.$$

For a fixed irrational number x, the order imposed by '$\prec$' is well-defined: given two rational numbers, one can definitely say which one precedes the other. So we may picture the rational numbers as arranged in an infinitely long list. The list extends in both directions; there is no 'first' fraction on the list, nor a last; one can always extend a given finite list both ways.

EXAMPLE. The above list may be extended to include 41/29 (preceding 7/5), and the modified list may, in turn, be extended to include 99/70 (preceding 41/29), and so on.

REMARK 3. The alert reader may notice that one can talk of an excellent approximation for a rational number too. For instance, for $x = 19/11$, the fractions 1/1, 2/1, 5/3, 7/4 are excellent approximations. However, the sequence of such fractions is finite: no improvement is possible beyond a point.

EXAMPLE. For $x = 19/11$, the sequence of excellent approximations is:

$$\frac{1}{1}, \quad \frac{2}{1}, \quad \frac{5}{3}, \quad \frac{7}{4}, \quad \frac{19}{11},$$

and no improvement is possible after this.

REMARK 4. As noted above, for any given irrational number x, the rational numbers may be placed in an ordered list. A remarkable property holds for these ordered lists: if a/b and c/d are consecutive fractions in any such list, then $ad - bc = \pm 1$. This is known as the "Farey" property, because it holds for the Farey sequences which are defined as follows. For any given positive number n, let $\mathcal{F}(n)$ be the sequence of distinct fractions p/q with $0 \leq p \leq q \leq n$ and $q > 0$, written in 'lowest form', with common factors cancelled away (i.e., $\gcd(p, q) = 1$), and arranged in ascending order. Thus:

$$\begin{aligned}
\mathcal{F}(1) &= \left\langle \frac{0}{1}, \frac{1}{1} \right\rangle, \\
\mathcal{F}(2) &= \left\langle \frac{0}{1}, \frac{1}{2}, \frac{1}{1} \right\rangle, \\
\mathcal{F}(3) &= \left\langle \frac{0}{1}, \frac{1}{3}, \frac{1}{2}, \frac{2}{3}, \frac{1}{1} \right\rangle, \\
\mathcal{F}(4) &= \left\langle \frac{0}{1}, \frac{1}{4}, \frac{1}{3}, \frac{1}{2}, \frac{2}{3}, \frac{3}{4}, \frac{1}{1} \right\rangle,
\end{aligned}$$

and so on. The property that $ad - bc = \pm 1$ for each pair of adjacent fractions a/b and c/d is easily verified.

⋆ ⋆ ⋆

It is easy to write a computer program to compute the sequence of excellent approximations corresponding to a given number. Here is the basic outline of the program (given in 'pseudo-code').

Step 0. Let `q` = 1; `bestdiff = 1`.

Step 1. `Input x`.

Step 2. Let `p` be the integer closest to `qx`; the fraction under consideration is now `p/q`. Compute the error (`diff`) by: `diff` = |`p` − `qx`|.

Step 3. If `diff` $<$ `bestdiff`, then print the fraction `p/q` and replace the current value of `bestdiff` by that of `diff`. (In computer language: `if diff < bestdiff then print p;"/";q : bestdiff := diff.`)

Step 4. Increment the value of `q` by 1 (i.e., `q` $=$ `q` $+$ `1`) and return to Step 2.

Here is the corresponding program written in BASIC, with `x` $= \sqrt{2}$.

```
REM excellent approximations
x = sqr(2): nmax = 1000: bestdiff = 1
for q = 1 to nmax
    p = int(q*x + 0.5) : diff = ABS(p - q*x)
    if diff < bestdiff then print p;"/";q; ";";
    if diff < bestdiff then bestdiff = diff
next q
```

The fractions obtained by running the program on a computer are displayed below.

$$\frac{1}{1}, \frac{3}{2}, \frac{7}{5}, \frac{17}{12}, \frac{41}{29}, \frac{99}{70}, \frac{239}{169}, \frac{577}{408}, \frac{1393}{985}, \ldots.$$

This sequence is rich in patterns; see if you can find and prove some of them. (Check that the Farey property holds.)

HINT. Study the numerators (N) and denominators (D) and check for relationships between N^2 and D^2. This should start you off on at least one family of patterns.

If we run the program with $x = \sqrt{10}$ and `nmax` $= 10000$ rather than 1000, we obtain another equally rich sequence:

$$\frac{3}{1}, \frac{19}{6}, \frac{117}{37}, \frac{721}{228}, \frac{4443}{1405}, \frac{27379}{8658}, \ldots.$$

Here are the fractions obtained when $x = \pi$:

$$\frac{3}{1}, \frac{22}{7}, \frac{355}{113}.$$

No further fraction is obtained even with `nmax` $= 10000$, which tells us that $355/113$ is the best rational approximation to π amongst the fractions with denominators less than 10000. The fraction is easy to remember: write '113355', slice the number into two halves, then arrange the pieces into a fraction. (With `nmax` $= 50000$, a new fraction is obtained: 104348/33215. This differs from π by approximately 3×10^{-10}.)

Run the program using other values for x:

- rational numbers (you will find that you obtain no new fraction from some point on);
- quadratic irrationals such as $\sqrt{5}$, $\sqrt{7}$, ...; the sequences here have an abundance of pretty patterns;
- higher degree irrationals such as $\sqrt[3]{2}$, $\sqrt[5]{10}$, ...

  (With these numbers the results are not so much inspiring as mystifying. There are, unfortunately, no simple patterns to be found in the sequences produced by higher degree irrationals.)

EXERCISES

2.2.1 Show that $1/1 \overset{\sqrt{2}}{\prec} 15/11 \overset{\sqrt{2}}{\prec} 2/1$.

2.2.2 Find a fraction p/q such that $2/1 \overset{\sqrt{2}}{\prec} p/q \overset{\sqrt{2}}{\prec} 5/4$.

2.2.3 Given a fraction a/b, show that another one may be found, say p/q, such that $p/q \overset{\sqrt{2}}{\prec} a/b$.

2.2.4 Find a fraction p/q such that

$$\frac{22}{7} \overset{\pi}{\prec} \frac{p}{q} \overset{\pi}{\prec} \frac{25}{8}.$$

(Here, the ordering is with reference to π, not $\sqrt{2}$.)

2.3 Logarithms

We shall begin by recalling the definition of a logarithm:

$$\text{if } b^c = N; \quad \text{then } \log_b N = c;$$

where b is the *base* of the logarithm. (Generally, we take b to be a positive integer greater than 1. However, any positive value distinct from 1 is permissible.) So, $\log_b N = c$ is just another way of writing $b^c = N$. Some examples are:

- $\log_{10} 100 = 2$ carries the same meaning as $10^2 = 100$;

- $\log_2 8 = 3$ carries the same meaning as $2^3 = 8$;
- $\log_4 2 = 1/2$ carries the same meaning as $4^{1/2} = 2$.

Logarithms were once an indispensable aid to computation, and countless navigators and pilots have staked their lives on them, not to mention engineers, astronomers and statisticians. Today, however, their role in computation is minimal, and the value of the logarithmic function lies chiefly in its rich mathematical properties.

However, the computational aspect can still be a source of fun! Consider the following table, which displays for a few numbers their logarithms to base 2.

x	1	2	4	8	16	32	$\cdots$
$\log_2 x$	0	1	2	3	4	5	$\cdots$

Observe that as x increases, so does $\log_2 x$. Naturally, we expect this to hold for intermediate values too; e.g.,

$$\text{if } 2 < x < 4 \text{ then } 1 < \log_2 x < 2,$$

and so on. More generally, if $0 < a < b$, then $\log_2 a < \log_2 b$.

For which x is $\log_2 x = 1.5$? The relation $\log_2 x = 1.5$ is equivalent to $x = 2^{1.5} = 2\sqrt{2}$, so the answer is $x \approx 2 \times 1.414$, or $x \approx 2.828$. Of course, we have 'cheated' here—we used our prior knowledge of the approximate value of $\sqrt{2}$.

What is the value of $\log_2 10$? Since $8 < 10 < 16$, it follows that $3 < \log_2 10 < 4$. It seems natural to expect that $\log_2 10$ is closer to 3 than to 4. In other words, that $3 < \log_2 10 < 3.5$, or (equivalently) $2^3 < 10 < 2^{3.5}$. This is so, and the simplest way to see why is by squaring all the numbers in the inequality; we reach the equivalent relation $2^6 < 100 < 2^7$, or $64 < 100 < 128$, which is true. Therefore, $3 < \log_2 10 < 3.5$, as claimed.

However, we can do better. Building up a table of powers of 2, we find that $2^{10} = 1024$. Also, $10^3 = 1000$, so we have $2^{10} \approx 10^3$; this is to be interpreted as saying that the ratio $2^{10}/10^3$ is close to 1. Therefore, taking roots, $10 \approx 2^{10/3}$, and so

$$\log_2 10 \approx \frac{10}{3} \approx 3.333, \qquad \log_{10} 2 \approx \frac{3}{10} = 0.3.$$

These are fairly good approximations; accurate computations (with a calculator once again—more cheating!) reveal that $\log_2 10 \approx 3.322$ and $\log_{10} 2 \approx 0.3010$, so the error in the estimate for $\log_2 10$

occurs only in the second decimal place (the percentage error is roughly 0.34%). This means that we have done rather well.

How do we find the approximate value of $\log_2 3$? We could start with the relation

$$2^3 = 8 \approx 9 = 3^2,$$

which shows that $3 \approx 2^{3/2}$ or $\log_2 3 \approx 1.5$. But this is a poor approximation; after all, 9/8 is not that close to 1. Can we do better? On computing the powers of 2 and 3, we find that 2^8 and 3^5 are somewhat close to one another ($2^8 = 256$, $3^5 = 243$, $256/243 \approx 1.0535$). We conclude that $\log_2 3 \approx 8/5$. This is still not very good, but it is superior to 3/2.

Can we do still better, armed with nothing more? We can, by using our earlier findings about $\log_2 10$. We proceed as follows: $2^3 = 8$ and $3^4 = 81$, $\therefore\ 3^4 \approx 2^3 \times 10^1$. Taking logarithms, $4\log_2 3 \approx 3 + \log_2 10$ or

$$4\log_2 3 \approx 3 + \frac{10}{3} = \frac{19}{10}.$$

Therefore, $\log_2 3 \approx 19/12$.

How good is this approximation? We use our 4-function calculator to compute the powers 2^{19} and 3^{12} and determine their ratio. Here is the result: $2^{19} = 524288$, $3^{12} = 531441$, $3^{12}/2^{19} \approx 1.0136$.

We have done quite well! Let us compare the results:

- 1st approximation: $\log_2 3 \approx \frac{3}{2} = 1.5$.
- 2nd approximation: $\log_2 3 \approx \frac{8}{5} = 1.6$.
- 3rd approximation: $\log_2 3 \approx \frac{19}{12} \approx 1.5833$.
- Actual value (using a calculator): $\log_2 3 \approx 1.58496$.

The error in the 3rd approximation is about 1%. Considering the meager ammunition used, our results are indeed impressive!

If we run the BASIC program listed earlier with $x = \log_2 3$ and `nmax` = 10000, we obtain the fractions listed below.

$$\frac{2}{1},\quad \frac{3}{2},\quad \frac{8}{5},\quad \frac{19}{12},\quad \frac{65}{41},\quad \frac{84}{53},\quad \frac{485}{306},\quad \frac{1054}{665}.$$

See if you can show that $\log_2 3 \approx 65/41$ without using a calculator or computer. (This implies having to show that $2^{65}/3^{41}$ is 'nearly' equal to 1. Using a calculator, we find that $2^{65}/3^{41} \approx 1.01153$, while $2^{84}/3^{53} \approx 0.997914$ and $2^{485}/3^{306} \approx 1.00102$.)

EXERCISES

2.3.1 Show, without using a calculator, that $\log_3 20$ lies closer to 2.75 than to 2.5.

2.3.2 Find good rational approximations for the following numbers:

(a) $\log_2 5$, $\log_2 6$, $\log_2 7$ and $\log_2 11$;
(b) $\log_3 5$, $\log_3 6$, $\log_3 7$ and $\log_3 10$.

2.3.3 Which is larger: $\log_4 8$ or $\log_5 10$?

2.3.4 Which is larger: $\log_8 16$ or $\log_{20} 50$?

2.3.5 Which is larger: $\log_5 9$ or $\log_7 13$?

⋆ ⋆ ⋆

COMMENT. During the course of the analysis above, we had to compute 2^{19}. How did you do the computation? Can it be done using less than 19 multiplications, using a 4-function calculator? (The answer is: *Yes.*) This makes for an interesting question: *What is the least number of multiplications needed to compute the number 2^{2001}?*

Chapter 3

Old-Fashioned Geometry

3.1 Appetizers

We are now in familiar territory! There are several reasons for including material from classical geometry, or Euclidean geometry as it is generally known. First, geometry offers a very wide range of possibilities as far as problem solving is concerned, in the form of theorems, riders and construction problems. Second, it offers a lot of scope for illustrating varied approaches in problem solving. Finally, in terms of age and antiquity, geometry is hard to beat!

The first few problems are well-known theorems.

EXERCISES

3.1.1 Prove that the internal angle bisectors of a triangle meet in a point (the *incentre* of the triangle).

3.1.2 Prove that the medians of a triangle meet in a point (the *centroid* of the triangle).

3.1.3 Prove that the altitudes of a triangle meet in a point (the *orthocentre* of the triangle).

3.1.4 **Alternate Segment Theorem** Points A, B, C lie on a circle Γ, and BT is tangent to Γ at B. Prove that angles CBT, BAC are either equal or supplementary. (How would you distinguish between the two cases?)

3.1.5 (Continuation) Two chords AB and CD of a circle Γ meet at E. Prove that $AE \cdot EB = CE \cdot ED$. (What happens if E lies outside Γ?)

3.1.6 (Continuation) A point P lies outside a circle Γ; PT is tangent to Γ at T, and PAB is a line (a *secant*) meeting Γ at A and B. Prove that $PA \cdot PB = PT^2$.

3.1.7 Formulate and prove the converses of Problems 3.1.4, 3.1.5 and 3.1.6. (See the comment below.)

⋆ ⋆ ⋆

Meaning of 'converse' Let p and q represent statements, or "propositions" to use the language of logicians. The converse of the assertion *"if p then q"* is the assertion *"if q then p"*. Thus, the 'if' and 'then' have, in a way, swapped roles.

EXAMPLE. Consider Pythagoras's theorem:

$$\textit{In } \triangle ABC, \textit{ if } \angle C = 90^\circ \textit{ then } a^2 + b^2 = c^2,$$

and its converse:

$$\textit{In } \triangle ABC, \textit{ if } a^2 + b^2 = c^2 \textit{ then } \angle C = 90^\circ.$$

Both statements are true. The converses asked for here need to be first formulated, then proved. Note that not all propositions have valid converses.

EXERCISES

3.1.8 (a) In $\triangle ABC$, let the internal bisector of $\angle BAC$ meet BC at D. Show that $AB/AC = BD/DC$.

(b) Show, conversely, that if D is a point on BC such that $AB/AC = BD/DC$, then AD bisects $\angle BAC$.

(c) What corresponding statement can be made about the external angle bisector?

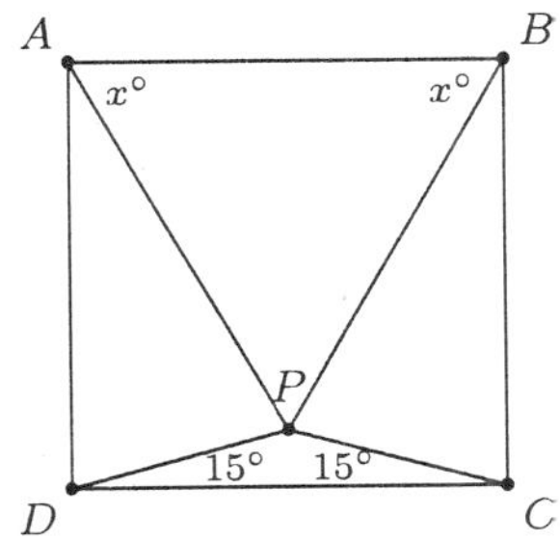

Figure 3.1. *Show that triangle PAB is equilateral!*

3.2 Proof by contradiction

In some of the problems in the preceding section you would probably have had to use "Proof by Contradiction"—a technique going back to the times of Euclid (known then as *Reductio ad Absurdum*). A few comments need to be made here regarding its usage. Essentially, the method rests on the principle that *self-contradiction is not acceptable.*

For instance, consider the following assertion: *There exists a largest integer.* This is clearly self-contradictory: if N is the supposedly largest integer, then the existence of the integer $N+1$ immediately brings forth a contradiction, because $N+1>N$. It follows that there can be no largest integer. This example may appear trivial (it is!), so we present a more convincing one.

Problem
Let $ABCD$ be a square, and let P be a point within it such that $\angle PCD = \angle PDC = 15°$. Show that $\triangle PAB$ is equilateral.
[See Figure 3.1.]

Solution Curiously, a direct proof by "angle-chasing" does not seem to work. Instead, we proceed as follows. Let angle PAB be $x°$. Since $PA = PB$, we have:

$$\angle PBA = x°, \quad \angle APB = (180-2x)°,$$
$$\angle PAD = (90-x)°, \quad \angle APD = (x+15)°.$$

Suppose that $x>60$. Then $3x>180$, $\therefore x>180-2x$, $\therefore \angle ABP > \angle APB$, $\therefore AP>AB$. Also, $x+15>75$, so in $\triangle APD$ we have $\angle APD > \angle ADP$, $\therefore AD>AP$. Thus, we have $AD>AP>AB$, an absurdity because $AD=AB$ ($ABCD$ is after all a square!).

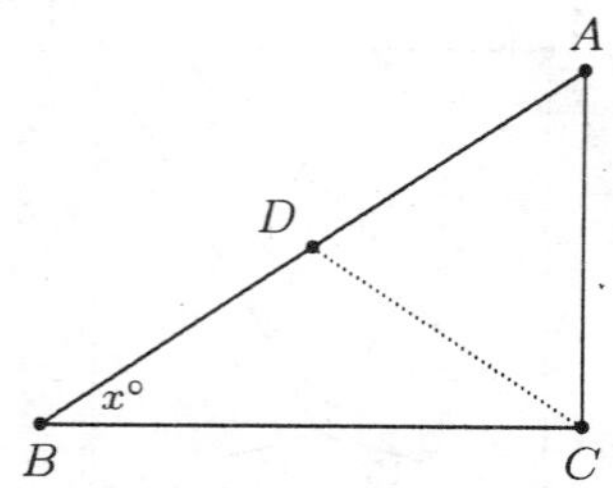

Figure 3.2. *Show that $DA = DC = DB$.*

So, we reach a self-contradiction and conclude that x cannot exceed 60.

What happens if we suppose that $x < 60$? By following the same line of reasoning, but with all the '>' signs changed to '<' and *vice versa*, we find that in this case too we are led to a self-contradiction. Therefore, x cannot be less than 60 either.

The only option left is to conclude that $x = 60$, and this means that $\triangle PAB$ is equilateral. □

The following challenge is left for the reader: *Solve the above problem without using proof-by-contradiction. In other words, find a "direct proof"!*

⋆ ⋆ ⋆

Here is another such example.

Problem

Let $\triangle ABC$ be right-angled with $\angle C = 90°$, and let the midpoint of the hypotenuse AB be D. Show that $DA = DC = DB$; that is, D is the circumcentre of $\triangle ABC$. [See Figure 3.2.]

Solution Let angle ABC be $x°$; then angle $BAC = (90 - x)°$.

We first suppose that $DB < DC$. Then $\angle DCB < x°$, $\therefore \angle DCA > (90 - x)°$ and so $\angle DCA > \angle DAC$, implying that $DA > DC$ and $DA > DC > DB$. But then we obtain $DA > DB$, and this is absurd because D is the midpoint of AB. We have at hand a self-contradiction.

The supposition is that $DB > DC$ leads to a similar contradiction. We conclude, then, that $DA = DC = DB$. □

There are of course other proofs available, and we challenge the reader to find a direct proof. A nice feature of the proof given above is that no construction lines were needed.

Further remarks on the nature of proof by contradiction will be found in the chapter on Number Theory.

EXERCISES

3.2.1 Prove the assertion made earlier in Chapter 1: *In* $\triangle ABC$, *if* $a^2 = b(b+c)$, *then* $\angle A = 2\angle B$.

3.2.2 Find a procedure for constructing $\triangle ABC$ given the lengths b, c and m_a (the length of the median through A).

3.3 Construction of triangles

Generally speaking, we require three independent pieces of data about a triangle in order to construct it: three sides; or two sides and the included angle; or two angles and one side; and so on. Sometimes, the pieces of information are not independent of one another; e.g., the three angles of the triangle (whose sum is 180°, so if we know any two of them, then we also know the third). A less obvious example is provided by the triple (R, r, d), where R and r are respectively the radii of the circumcircle and incircle of the triangle and d is the distance between the two centres; the relation connecting them is $d^2 = R(R-2r)$. In such cases, the 'third' piece of information has no value, so the given information is insufficient to construct the triangle. Expressed otherwise, there are infinitely many triangles that share the given specifications. This may remind you about certain problems in algebra; e.g., problems such as *"Find* x, y *given that* $x+2y = 6$ *and* $3x+6y = 18$*"*, which have infinitely many solutions. (In this example, the 2nd equation follows from the 1st one and therefore carries no new information.)

In some cases, there may be two or more (but not infinitely many) triangles that fit the given data; this would be analogous to there being two solutions for a quadratic equation, both relevant to the problem at hand. Finally, in some cases, it may happen that there is no solution at all; i.e., no triangle can be found that fits the given data. Once again, there are obvious analogies from algebra; e.g., the problem *"Find* x, y *given that* $x + 2y = 6$ *and* $3x + 6y = 11$*"* which clearly has no solution, because the two equations contradict one another.

A note on notation We list below some symbols in common usage.

- a, b, c are the lengths of the sides BC, CA, AB of $\triangle ABC$.
- m_a, m_b and m_c are the lengths of the medians through A, B and C.
- t_a, t_b and t_c are the lengths of the angle bisectors through A, B and C.
- h_a, h_b and h_c are the lengths of the altitudes through A, B and C.
- r is the radius of the incircle of the triangle.
- R is the radius of the circumcircle of the triangle.

EXERCISES

3.3.1 Give an example of a (b, c, m_a) construction problem for which no solution exists.

3.3.2 (Continuation) Carry forward the analysis and find a valid continuation of the following sentence:
A (b, c, m_a) triangle exists if and only if ....

3.3.3 Make a complete analysis of the (b, c, h_a) construction problem.

3.3.4 Show that for the construction problem (b, c, R) there are, in general, two solutions. Determine the conditions under which there exists: (i) precisely one solution, and (ii) no solution.

3.3.5 In general, how many solutions are there for the following construction problems?

(a) (b, c, h_b);
(b) (h_a, h_b, h_c);
(c) (A, t_a, r);
(d) (A, t_a, c);
(e) (a, c, m_a).

★★★

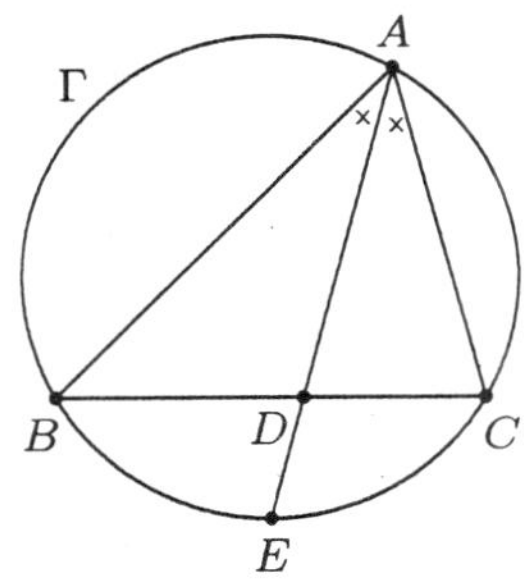

Figure 3.3. *Formula connecting b, c and t_a*

Traditionally, we are encouraged to use only 'pure geometry' when we tackle problems in geometry. The author however holds no such ideology, and instead encourages his readers to use any idea and any lead whatsoever that seems relevant to the problem at hand. Elegance is certainly a necessary ingredient in mathematics—but that comes later on! To illustrate what this means, consider the (b, c, t_a) construction problem. We shall show *algebraically* that the specifications serve to fix $\triangle ABC$ uniquely.

We first show how to compute t_a given a, b, c. Let $\triangle ABC$ have circumcircle Γ (see Figure 3.3), and let the internal bisector of $\angle A$ meet BC at D and Γ again at E. The following facts are available:

- $\dfrac{AB}{AC} = \dfrac{BD}{DC}$;
- $AD \cdot DE = BD \cdot DC$;
- $\triangle ABE \sim \triangle ADC$ (for $\angle BAE = \angle DAC$, $\angle ABD = \angle AEC$).

The first two relations yield

$$BD = \frac{ac}{b+c}, \quad DC = \frac{ab}{b+c}, \quad AD \cdot DE = \frac{a^2\,bc}{(b+c)^2}.$$

From the third relation, we obtain $AB/AE = AD/AC$, or $AD \cdot AE = AB \cdot AC = bc$. Subtraction now yields

$$t_a^2 = AD^2 = AD \cdot AE - AD \cdot DE = bc\left(1 - \frac{a^2}{(b+c)^2}\right).$$

We now have at hand the desired formula giving t_a in terms of a, b, c. We deduce that

$$\frac{t_a^2}{bc} + \left(\frac{a}{b+c}\right)^2 = 1,$$

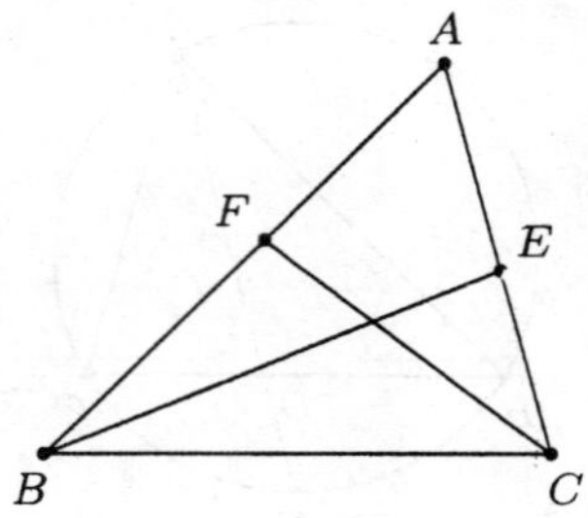

Figure 3.4. *The Steiner–Lehmus problem*

and this shows that once b, c, t_a are known, a is fixed; indeed,

$$a = (b+c)\sqrt{1 - \frac{t_a^2}{bc}}.$$

Thus, the specifications (b, c, t_a) serve to fix $\triangle ABC$ uniquely.

⋆ ⋆ ⋆

This analysis yields an unexpected bonus—an algebraic solution to the following notorious problem.

> **Steiner–Lehmus Problem.** *In a triangle ABC let the internal bisectors of angle B and angle C meet the opposite sides (AC and AB) at E and F respectively. Suppose that $BE = CF$. Show that $AB = AC$.* [See Figure 3.4.]

The problem looks simple—but this is deceptive! Indeed, it has tested the ingenuity of many generations of mathematicians. We present here a solution using algebra. Clearly, the problem may be stated as follows:

In $\triangle ABC$, if $t_b = t_c$, then $b = c$.

Now, the condition $t_b = t_c$ is equivalent to the following:

$$ac\left(1 - \frac{b^2}{(a+c)^2}\right) = ab\left(1 - \frac{c^2}{(a+b)^2}\right).$$

On clearing the fractions and cancelling away the common factor $a(a+b+c)$, we obtain

$$c(a+b)^2(a+c-b) = b(a+c)^2(a+b-c).$$

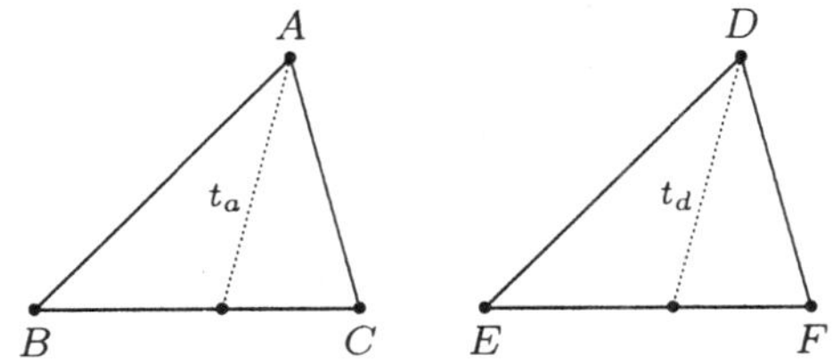

Figure 3.5. *A criterion for congruence.*

Bringing all the terms to one side, we find that $(c-b)$ is a divisor of the resulting expression. Factorization yields:

$$(c-b)\left(a^3+a^2(b+c)+3abc+b^2c+bc^2\right)=0.$$

(Derive was used to find the factorization!) Since a, b, c are positive, the quantity

$$\left(a^3+a^2(b+c)+3abc+b^2c+bc^2\right)$$

cannot be 0. It follows that $c-b=0$, or $b=c$. We have proved the Steiner–Lehmus assertion! □

⋆ ⋆ ⋆

It would not be out of place here to offer a 'pure geometry' proof of this result. We shall first prove the following theorem.

> **Theorem (A criterion for congruence)** *Let $\triangle ABC$ and $\triangle DEF$ be such that $a=d$, angle A equals angle D, and $t_a=t_d$. Then, $\triangle ABC \cong \triangle DEF$.* [See Figure 3.5.]

For proof, we note that if we place the triangles so that BC and EF coincide (with E on B and F on C) and A and D lie on the same side of BC, then the circumcircles of the two triangles coincide; for the circles have equal radius (equal to $a/\sin A$). Let Ω denote the common circumcircle (see Figure 3.6).

Let M be the midpoint of the arc BC not containing A, and let ℓ be the diameter of Ω through M. By reflecting through ℓ, if necessary, we may assume that A and D lie on the same side of ℓ as C; i.e., $\angle B \le \angle C$ and $\angle E \le \angle F$, as shown in Figure 3.6. Let AM and DM (the bisectors of angles BAC and EDF) meet BC at P and Q, respectively. Then, $t_a=AP$ and $t_d=DQ$.

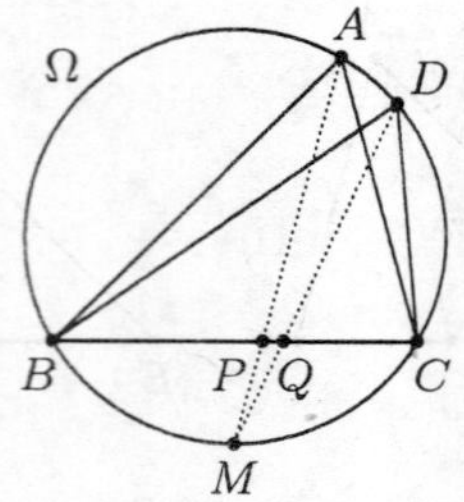

Figure 3.6. *Proof of the criterion*

If A and D coincide, then there is nothing to prove. If not, we suppose without loss of generality that $\angle E < \angle B$; that is, the points B, A, D, C occur in that order on arc BC. In this case, $DM < AM$ and $QM > PM$ (for $\angle C > \angle B > \angle E$, therefore $\angle MPQ = \angle A/2 + \angle C > \angle D/2 + \angle E = \angle MQP$). But then $DQ < AP$, since $DQ = DM - QM$ and $AP = AM - PM$. This contradicts the hypothesis that $t_a = t_d$.

It follows that the points A and D coincide. That is, $(a, \angle A, t_a) = (d, \angle D, t_d)$ implies $\triangle ABC \cong \triangle DEF$. □

COMMENT. This proof provides another instance of proof by contradiction.

The criterion allows an easy proof of the Steiner–Lehmus result. Consider $\triangle ACF$ and $\triangle ABE$ (see Figure 3.4). The two triangles share the same apex angle ($\angle A$), have equal bases ($CF = BE$) and equal length of the bisector of the apex angle (AI, where I is the incentre of $\triangle ABC$). It follows that $\triangle ACF \cong \triangle ABE$, and so $AC = AB$. Thus, the triangle is isosceles, as claimed. □

⋆ ⋆ ⋆

A more impressive example of the power of algebraic methods is provided by the following analysis. We shall show that the specifications (R, r, s) suffice to uniquely fix a triangle; in other words, that the lengths of the sides can be found from R, r and s.

Let the triangle be denoted by ABC. We use the following known relations:

(a) $2R = \dfrac{a}{\sin A} = \dfrac{b}{\sin B} = \dfrac{c}{\sin C}$;

(b) $r = (s-a)\tan\dfrac{A}{2} = (s-b)\tan\frac{1}{2}B = (s-c)\tan\frac{1}{2}C$;

(c) $\sin A \tan \dfrac{A}{2} = 2\sin^2 \dfrac{A}{2}, \qquad \sin A \cot \dfrac{A}{2} = 2\cos^2 \dfrac{A}{2}.$

From (c) we obtain the identity:

$$\sin A \left(\tan \frac{A}{2} + \cot \frac{A}{2} \right) = 2 \text{ for all angles } A.$$

Combining this with (a) and (b), we deduce that

$$\frac{a}{2R} \left(\frac{r}{s-a} + \frac{s-a}{r} \right) = 2.$$

Clearly, if b or c are substituted for a we shall still have valid identities. This means that a, b and c are roots of the following equation in x:

$$\frac{x}{2R} \left(\frac{r}{s-x} + \frac{s-x}{r} \right) = 2.$$

On simplification, this yields the following equation:

$$x^3 - 2sx^2 + (r^2 + 4rR + s^2)x - 4rsR = 0.$$

This is a *cubic* equation, and it can have at most three real roots. Since a, b and c satisfy the equation, these must be the only roots of the equation. Since the equation is 'fixed' by (R, r, s), it follows that the sides of the triangle are also fixed by (R, r, s); i.e., the triangle is fully determined. This proves our claim. □

⋆ ⋆ ⋆

It will be clear that the list of construction problems can be lengthened considerably! Some of these problems are quite challenging, e.g.:

(a) Construct $\triangle ABC$ given (b, c, t_a).

(b) Construct $\triangle ABC$ given (a, r, R).

However, we shall not pursue these questions further, as it would involve a much more detailed study of geometry than is intended to be done here. The interested reader should refer to some of the excellent texts available—the texts by C V Durell, S H M Coxeter & S Greitzer, Max Jeger, I M Yaglom, R A Johnson, and others.

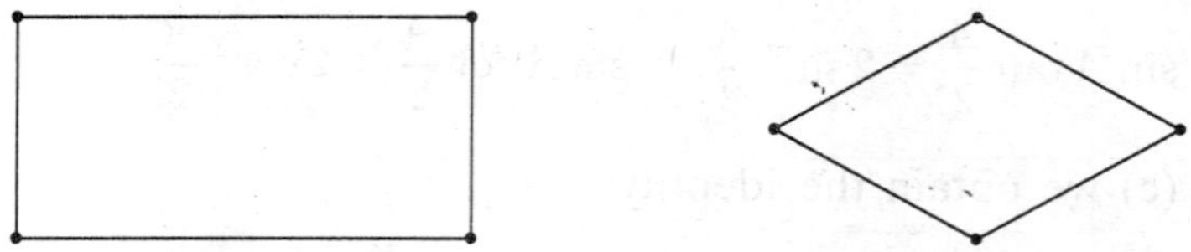

A rectangle: equiangular but not equilateral

A rhombus: equilateral but not equiangular

Figure 3.7. *Equal sides, equal angles*

3.4 Equal angles, equal sides

Occasionally, one comes across the following definition for an equilateral triangle: *An equilateral triangle is one that has equal sides and equal angles.* But this is clearly flawed, for the two conditions ('equal sides' and 'equal angles') are equivalent to one another: each one implies the other. So the definition contains redundant information.

What happens for polygons with more sides? We refer to a polygon as *equilateral* if it has equal sides, and *equiangular* if it has equal angles. It is easy to see that when n, the number of sides, exceeds 3, the two conditions are no longer equivalent to one another. For example, consider a non-square rectangle; here $n = 4$ and 'equal angles' holds but not 'equal sides'. For a non-square rhombus, the situation is reversed.

Similar statements can clearly be made for $n > 4$. Starting with a regular pentagon, one can move a side parallel to itself; this will result in an equiangular but non-equilateral pentagon. Or, treating the original shape as an assemblage of rods of equal length, one can compress the shape, pushing inwards from two opposite ends. As the object is non-rigid, the resulting shape will be equilateral but non-equiangular. The situation now starts to look unclear but positive statements can still be made.

EXERCISES

3.4.1 In the case of the pentagon, prove that if it is known to be *cyclic* (i.e., its vertices lie on a circle), then equal angles imply equal sides, and conversely.

3.4.2 For the case of the cyclic hexagon, what is the most that can be said?

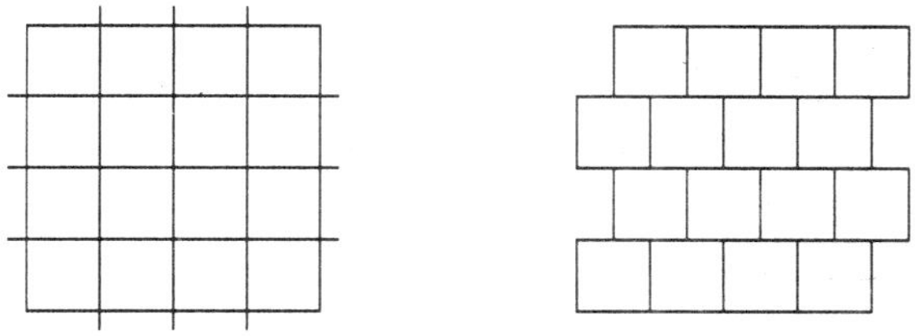

Figure 3.8. *Two possible tilings with squares.*

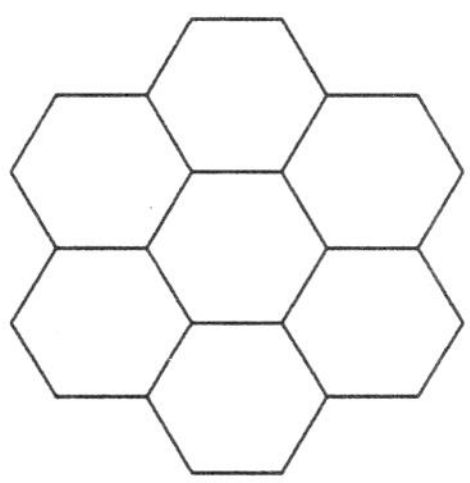

Figure 3.9. *Tiling with regular hexagons.*

3.4.3 What can be said for the case of a cyclic heptagon?

3.4.4 What generalization is possible from these results?

3.5 Tiling the plane

Imagine that one had infinitely many copies of a given polygon $\mathcal{P}$; the copies are all congruent to one another and their supply is unlimited. What properties should $\mathcal{P}$ have if, by using the copies, one can 'tile' the whole plane?—i.e., fill the plane with them so that *there are neither overlaps nor gaps left*. Such a tiling is certainly possible if $\mathcal{P}$ is a square (and it can be done in several ways; see Figure 3.8). A tiling is similarly possible using a rectangle, and it is also possible using a regular hexagon (see Figure 3.9). This is the familiar beehive pattern; you may have seen it used in bathroom flooring.

EXERCISES

3.5.1 With which triangles can the plane be tiled?

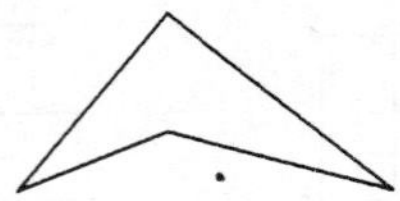

Figure 3.10. *A boomerang*

3.5.2 With which convex quadrilaterals can the plane be tiled? (Consider the elementary cases first: parallelograms, kites, rhombuses, trapezoids, ...)

3.5.3 [*] What happens if the quadrilateral is non-convex? (A non-convex or re-entrant quadrilateral is that in which one of the internal angles exceeds 180 degrees; see Figure 3.10. Such shapes are also known as *boomerangs*.)

Can a convex polygon be tiled using only boomerangs? That is, can a finite number of boomerangs be arranged so as to make a convex polygon? (Note: The boomerangs are not required to be identical copies of one another.)

3.5.4 Can the plane be tiled using regular pentagons?

3.5.5 Find a non-regular pentagon with which the plane can be tiled.

HINT. The task is simplified by restricting the search to pentagons that have an axis of symmetry.

3.5.6 Find a non-regular hexagon with which the plane can be tiled.

3.5.7 Some polygons have the property that by taking a certain (finite) number of copies of them, one can fit the copies together without overlaps or gaps to obtain a polygon similar in shape to the original one (obviously of a larger size). Clearly, one can do this with any rectangle or parallelogram, with four pieces used in each case (see Figure 3.11). Shapes having such a property are called *reptiles* (for 're-peating tiles').

(a) Show that every triangle is a reptile.

(b) Construct a reptile which is trapezoidal but *not* a parallelogram.

(c) Construct a non-trapezoidal quadrilateral reptile.

(d) What can be said in the case of pentagons?

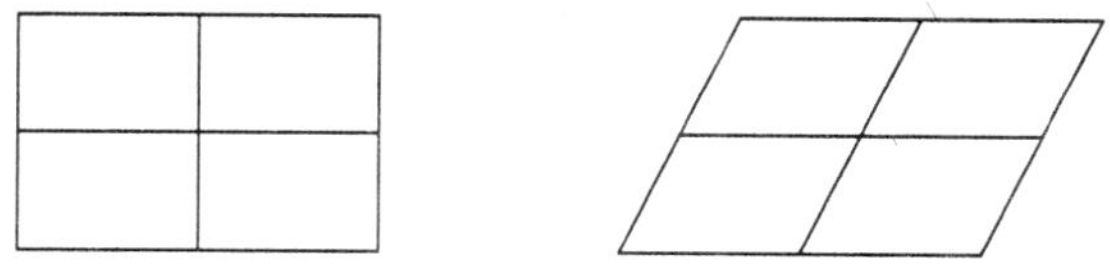

Figure 3.11. *Two reptiles.*

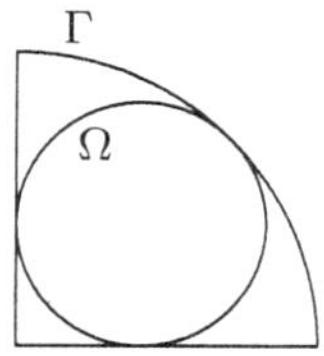

Figure 3.12. *Circle in a quarter-circle*

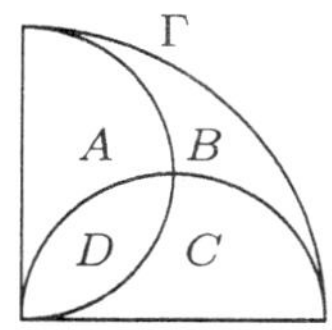

Figure 3.13. *Two semi-circles in a quarter-circle*

3.6 Wheels within wheels

A rich variety of problems becomes available when we consider configurations of circles that touch each other. Remarkably, the methods of proof are all very similar and draw upon just one basic result: that the centres of two circles that touch one another are collinear with the point of contact. This fact, used in conjunction with Pythagoras's theorem, yields a surprising amount.

EXERCISES

3.6.1 Figure 3.12 shows a circle, Ω, symmetrically inscribed in a quarter-circle Γ of radius r. Find the radius of Ω.

3.6.2 Within a quarter-circle Γ are drawn two semi-circles as shown (Figure 3.13); and four regions A, B, C, D are produced. Find the ratios of their areas, i.e.,

$$\text{Area}(A) : \text{Area}(B) : \text{Area}(C) : \text{Area}(D).$$

3.6.3 Let $\triangle ABC$ be right-angled, with $\angle A = 90°$, and let its circumcircle be Γ (see Figure 3.14). Let Ω be the circle which touches sides AB and AC and circle Γ (internally). Show that the radius r' of Ω is given by

$$r' = \frac{2bc}{a+b+c},$$

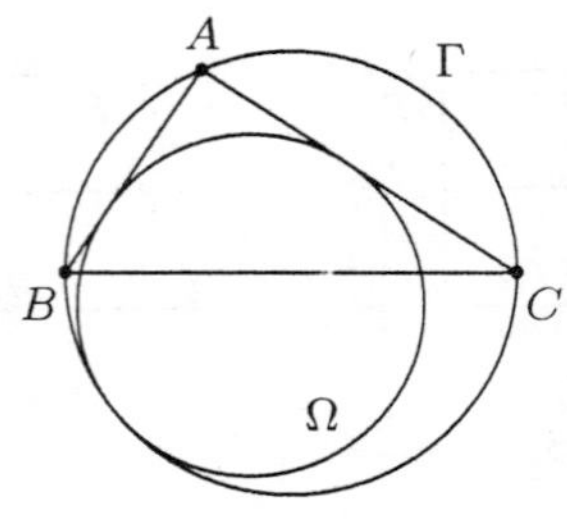

Figure 3.14. *Circle in a right-angled triangle*

and deduce, as a corollary, that the radius of Ω is *twice* that of the incircle of the triangle.

3.6.4 $\boxed{*}$ (Generalization)

Let $\triangle ABC$ have a circumcircle Γ, and let Ω be the circle that touches the sides AB and AC and also touches Γ internally. Let the points of contact of Ω with AB and AC be P and Q, respectively, and let I be the incentre of $\triangle ABC$. Show that I is the midpoint of PQ.

How is this problem a generalization of the previous one?

3.6.5 **The Shoemaker's Knife** On a straight line ℓ are marked points A, B and C (in that order). Semi-circles Γ_1, Γ_2 and Γ_3 are drawn on AB, BC and AC as diameters, all lying on one side of ℓ. Next, a chord DB of semi-circle Γ_3 is drawn perpendicular to the diameter AC (see Figure 3.15).

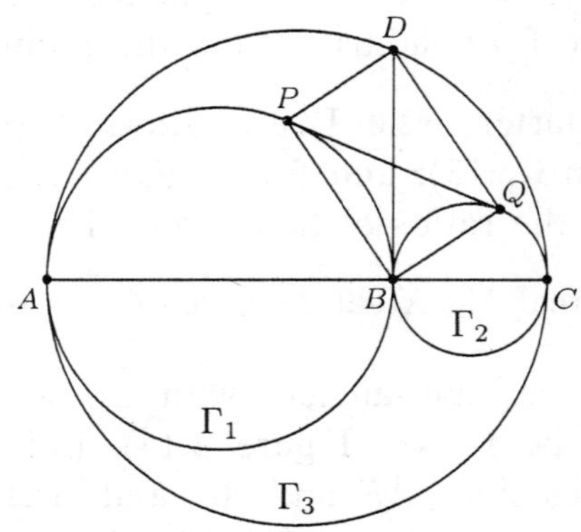

Figure 3.15. *The shoemaker's knife–1*

(a) The segment PQ is drawn tangent to Γ_1 and Γ_2 at P and Q, respectively (see Figure 3.15; full circles have been drawn rather than semi-circles, to avoid a visual clutter). Show that $BPDQ$ is a rectangle.

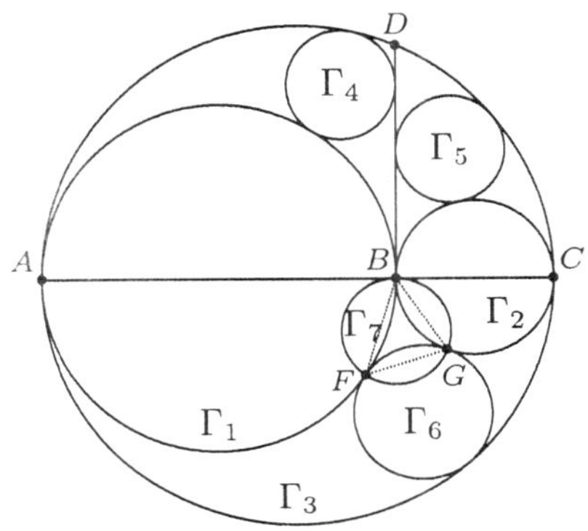

Figure 3.16. *The shoemaker's knife–2*

(b) Circles Γ_4 and Γ_5 are drawn as follows: Γ_4 touches Γ_3 (internally), Γ_1 (externally) and line DB; Γ_5 touches Γ_3 (internally), Γ_2 (externally) and line DB (so Γ_4 and Γ_5 lie on opposite sides of the line DB (see Figure 3.16). As earlier, full circles have been drawn rather than semi-circles, to avoid clutter). Show that circles Γ_4 and Γ_5 have equal radius.

⋆ ⋆ ⋆

COMMENT. The equality of radii of the Γ_4, Γ_5 is one of the very many remarkable facts about this configuration (which is historically known as the *shoemaker's knife*; to see why, hide certain portions of the diagram). The reader will be pleasantly surprised to learn that a new result about this configuration was discovered as recently as in 1979. Here is the result (due to Leon Bankoff, a dentist and amateur mathematician in the USA).

We start by drawing circles Γ_1, Γ_2, Γ_3, Γ_4, Γ_5 as earlier. Next, we draw the circle Γ_6 which touches Γ_1 and Γ_2 (externally) and Γ_3 (internally). (To avoid a messy figure, draw Γ_6 on the 'other' side of ℓ, as in Figure 3.16.) Let the points of contact of circle Γ_6 with Γ_1 and Γ_2 be F and G, respectively. Finally, we draw Γ_7, the circumcircle of triangle BFG. Then, circles Γ_7, Γ_4 and Γ_5 are congruent to one another!

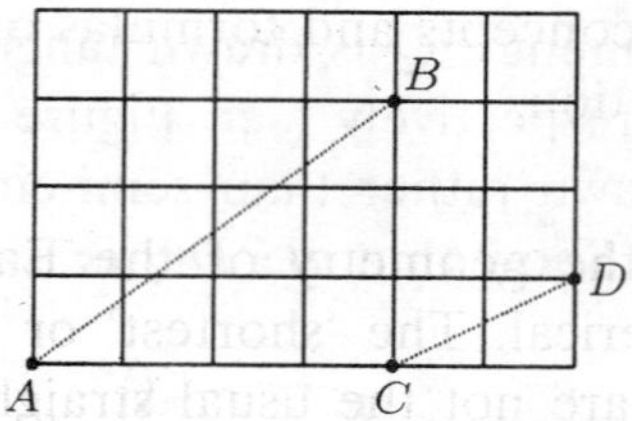

Figure 3.17. *Taxicab geometry.*

EXERCISES

3.6.6 Prove Bankoff's Theorem.

3.7 Taxicab geometry

Suppose we live in a city where the roads run only North–South or East–West, with the roads spaced out at regular intervals as on a rectangular grid. Such cities actually exist—to a first approximation, at any rate; portions of Mumbai, Bangalore and Manhattan Island in New York City are constructed in this manner. Let a coordinate system be imposed upon the city.

Consider the points $A\,(0,0)$ and $B\,(4,3)$, shown in Figure 3.17. Let distances be referred to by the symbol d. What is the distance $d(A,B)$ between A and B?

Pythagoras's theorem tells us that $d(A,B) = 5$. But is this a realistic measure of distance? Can we really walk or drive along the segment AB? To do so, we may have to march through various private residences! If the roads all run E–W or N–S, then perhaps the best we can do is to travel 4 units East and then proceed 3 units North (or 3 units North followed by 4 units East; the order does not matter), thereby traversing a total of 7 units. In a sense, this *is* the distance between A and B.

Likewise, the distance between points $C\,(4,0)$ and $D\,(6,1)$ is $2+1=3$ units. More generally, we take the distance between $A\,(a,b)$ and $B\,(c,d)$ to be given by the formula $|a-c|+|b-d|$. (Earlier, it was given by the Pythagorean formula: $\sqrt{(a-c)^2+(b-d)^2}$.)

⋆ ⋆ ⋆

Note that we have, in effect, changed the formula for distance. Changing so basic a formula may seem a rather drastic step to take, but in modern mathematics this happens routinely. It must be remembered that concepts and formulas often do get modified to meet a given situation.

EXAMPLE. Consider the geometry of the Earth's surface, which we may take as spherical. The 'shortest' or 'straightest possible' lines on this surface are not the usual straight lines; indeed, it is not possible to draw a straight line upon a spherical surface. It can be shown that the shortest lines (*geodesics*, as they are called) are arcs of great circles; i.e., circles with the same centre as the sphere. Airline routes follow great circle paths for precisely this reason. (For instance, the route from Rome to Chicago goes over Greenland.) So, to measure the distance between two points, we measure the arc length along the great circle passing through the two points. (There is precisely one great circle passing through two given points on the sphere, except when the two points are antipodes of one another, in which case there are infinitely many such circles.) The distance formula now is not the usual Pythagorean one, but one that depends in a rather complicated way on the latitude and longitude of the two points. This formula has been known to mathematicians for a long time and is of great importance to navigators. (The Pythagorean formula would give us the lengths of straight-line tunnels through the Earth!)

⋆ ⋆ ⋆

To avoid confusion between the two formulae, we shall use the terms *taxicab distance* and *Euclidean distance*; the corresponding symbols will be d_T and d, respectively. Thus, given two points $A\,(a,b)$ and $B\,(c,d)$, we have

$$\begin{aligned} d_T(A,B) &= |a-c|+|b-d|, \\ d(A,B) &= \sqrt{(a-c)^2+(b-d)^2}. \end{aligned}$$

Once the formula of distance has been modified, it is natural to expect that other geometrical notions will need to be worked out afresh; 'natural' because geometry as we know it—'Euclidean geometry'—is largely based on the notion of distance. Recall that concepts such as congruence and similarity, and definitions such as that of a circle, all ultimately rest on the notion of distance. If we alter the formula of distance, then the geometry gets altered in an essential manner; we obtain a new version of geometry! We refer to the new version as *taxicab geometry*.

To study taxicab geometry further, we shall articulate some questions in the form of problems (below). The discussion throughout is with reference to points on some fixed plane.

EXERCISES

3.7.1 In Euclidean geometry, a circle with centre O and radius r is defined to be the set of points P in the plane with $d(O,P) = r$. In taxicab geometry, we define a *taxicab circle* with centre O and radius r to be the set of points P in the plane with $d_T(O,P) = r$.

Question. What 'shape' does a taxicab circle have; how 'circular' is it? What symmetries does it possess?

3.7.2 In Euclidean geometry, the set of points equidistant from two fixed points A and B is a straight line bisecting segment AB at right angles. What happens in taxicab geometry—that is, what is the shape of the locus:

$$\{P : d_T(A,P) = d_T(B,P)\}$$

for given points A and B?

3.7.3 In Euclidean geometry, there is a unique shortest path connecting any two given points. In taxicab geometry, however, there can be a multiplicity of such paths. This has an unexpected implication on the concept of perpendicularity.

In Euclidean geometry, it is possible to define the notion of perpendicularity via that of distance (thus, reversing the normal development). We start with the Pythagorean formula, treating it as an axiom. Next, we define perpendicularity: if ℓ is a line and A and B are points with $A \notin \ell$ and $B \in \ell$, then AB is said to be perpendicular to ℓ if among all points on ℓ, B is the closest possible from A; that is,

$$d(A,B) \leq d(A,P) \text{ for all } P \in \ell.$$

This mode of developing the subject is not as eccentric as it might seem at first; the results obtained are identical to those obtained more traditionally.

In taxicab geometry, we define perpendicularity in the same way: given a line ℓ and a point $A \notin \ell$, we locate the

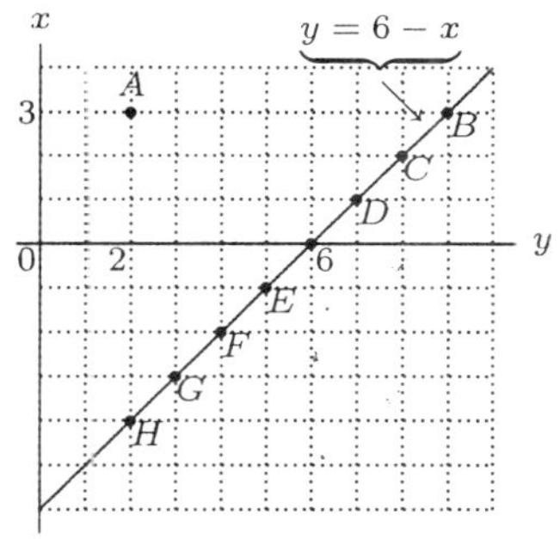

Figure 3.18. *Multiple perpendiculars!*

point(s) $B \in \ell$ such that $d_T(A, B)$ has the least possible value, and then we declare AB to be perpendicular to ℓ.

Here is a rather curious consequence of this definition. Let ℓ be the line $y = x - 6$ and A the point $(2, 3)$. Then, the points $B\,(9,3)$, $C\,(8,2)$, $D\,(7,1)$, ..., $G\,(3,-3)$, $H\,(2,-4)$ on ℓ are all at the least possible distance, namely 7, from A. (See Figure 3.18.) So here AB, AC and AD are all "perpendicular" to ℓ!

Classify all the possible situations where this phenomenon can occur in taxicab geometry.

3.7.4 In Euclidean geometry, the inequality

$$d(A, C) \leq d(A, B) + d(B, C)$$

holds for any three points A, B, C, and equality holds if and only if B lies on segment AC. This is the *triangle inequality*, and it is of very great importance in Euclidean geometry.

Does the inequality hold in taxicab geometry? That is, for arbitrary points A, B and C, do we always have $d_T(A, C) \leq d_T(A, B) + d_T(B, C)$? If equality holds, can we conclude that A, B, C are collinear (with B between A and C)?

(We need to do more than merely check a few instances; a general treatment is required.)

3.7.5 Let A and B be fixed points in the plane, and let k denote a positive constant. In Euclidean geometry, the locus

$$\{P : d(P, A) + d(P, B) = k\}$$

is an *ellipse*. For the locus to be non-empty, we must have $k \geq d(A, B)$. If $k = d(A, B)$, then the locus degenerates to the line segment AB.

What shape does the 'taxicab ellipse'

$$\{P : d_T(P, A) + d_T(P, B) = k\}$$

have? Here, $k \geq d_T(A, B)$.

3.7.6 [*] In this problem we study *isometries of the plane.* A transformation f of the plane is known as an *Euclidean isometry* if it preserves distance; that is, if

$$d(A, B) = d(f(A), f(B)) \text{ for all points } A \text{ and } B.$$

Reflection in a line is an example of a Euclidean isometry, as is rotation about a point (through any angle); these transformations clearly preserve distance.

It is an instructive and non-trivial exercise to classify all such isometries. It turns out that the following constitutes a complete list:

- reflections (in all possible axes);
- rotations (about all possible points, and by all possible angles);
- displacements (in all possible directions, and by all possible amounts); and
- arbitrary combinations of the above three operations.

The only isometry that does not belong to any of the "pure" classes is the *glide reflection*, which is a composite of a displacement and a reflection.

Make a complete list of the *taxicab isometries.* (It turns out that the list is quite small.)

3.7.7 Geodesics on a Curved Surface.
Identify the shortest lines: (a) on a cylindrical surface, and (b) on a conical surface.

3.7.8 (Continuation) Give an intuitive argument to justify the claim that the shortest lines on a spherical surface are arcs of great circles.

⋆ ⋆ ⋆

It is clear that many more such questions can be explored in taxicab geometry—the notions of congruence, similarity, and so on. A natural question is: *How can the concept of angle be defined meaningfully?* Here is one possible approach.

In Euclidean geometry, a triangle may be checked for acute-, right- or obtuse-angledness as follows. Let $\triangle ABC$ have sides a, b and c; then $\angle A$ is less than, equal to, or greater than 90° in accordance with whether $b^2 + c^2 - a^2$ is positive, zero or negative; and similarly for the other two angles. This suggests an approach by which angles can be brought in via the formula for distance. Thus, given $\triangle ABC$, we define the angle at A to be given by the formula

$$\cos A = \frac{b^2 + c^2 - a^2}{2bc}, \quad 0^\circ \leq A \leq 180^\circ.$$

The triangle inequality ($|b - c| < a < b + c$) ensures that this formula yields a meaningful value for $\angle A$.

In taxicab geometry, we proceed the same way: given three points A, B and C, let a, b and c denote the lengths $d_T(B, C)$, $d_T(C, A)$ and $d_T(A, B)$, respectively. Then, $\angle A$ is given by

$$\cos A = \frac{b^2 + c^2 - a^2}{2bc}, \quad 0^\circ \leq A \leq 180^\circ.$$

As earlier, this yields a meaningful value for $\angle A$.

EXAMPLE. Consider the three points $A = (0, 0)$, $B = (2, 3)$ and $C = (1, 6)$. The taxicab lengths of the sides of triangle ABC are: $a = 4$, $b = 7$, $c = 5$. We now have

$$\cos A = \frac{7^2 + 5^2 - 4^2}{2 \cdot 7 \cdot 5} = \frac{58}{70} = 0.8286;$$

therefore, $\angle A = \cos^{-1} 0.8286 \approx 34.05^\circ$.

The consequences of this remain to be explored, and we invite the reader to do so.

3.8 Minkowski geometry

As stated earlier, the exploration embarked upon above (of 'Taxicab Geometry') is typical of modern mathematics. We shall consider in this section another such example. Suppose that the distance between the points (a, b) and (c, d) is given neither by the Pythagorean nor the taxicab formula, but instead by

$$\sqrt{-(a - c)^2 + (b - d)^2}.$$

Observe that the formula is not symmetric with respect to the x- and y-coordinates.

This choice of formula gives rise to yet another version of geometry, now known as *Minkowski Geometry*—in honour of Hermann Minkowski who did an intensive study of this geometry as part of an attempt to place Einstein's Special Theory of Relativity on a geometric footing. Einstein published his theory in 1905, and in 1908 Minkowski published a paper titled *Space and Time*, which begins with the following very beautiful passage (it has subsequently become very famous):

> The views of space and time which I wish to lay before you have sprung from the soil of experimental physics, and therein lies their strength. They are radical. Henceforth space by itself, and time by itself, are doomed to fade away into mere shadows, and only a kind of union of the two will preserve an independent entity.

The thesis that Minkowski develops is that space and time need to be viewed as different coordinates of a 4-dimensional system, in which distances are measured by a formula similar to the one above. Once this is done, many of the spectacular predictions of special relativity follow quite naturally: contraction of length, dilatation of time, ... Minkowski stresses that this view of space–time is not a mere fiction but is forced upon us by the observed behaviour of light. This becomes evident when we study the results of experiments conducted in the 1890s by Michelson and Morley to measure the absolute velocity of the Earth in its movement through space. (The baffling result found by Michelson–Morley was that the velocity of light seemed entirely unaffected by the movement of the Earth. *Conclusion:* Either the Earth has zero velocity, and therefore occupies a privileged position in the universe—but this is absurd; or light behaves in a way quite contrary to common sense. It was Einstein who showed the way out of this impossible situation.)

An immediate observation about Minkowski geometry is that it is possible for distances to be imaginary. For example, the distance between $(1,0)$ and $(0,0)$ is the imaginary number $\sqrt{-1}$ (usually denoted by i). This may sound absurd and fantastic, something from science fiction rather than mathematics. Another consequence is that for each point A, there are infinitely many points P distinct from A and at zero distance from A. For instance, the distance between $(1,1)$ and $(2,2)$ is zero.

In the General Theory of Relativity, which deals with the effects of gravitation on space–time, the formula for distance undergoes further distortions. Historically, the first rigorous study of such 'distortions' was carried out by Bernhard Riemann in the 1850s, and it required another brilliant insight, this time on Einstein's part, before Riemannian geometry could be made use of in the study of space–time curvature produced by a gravitational field. It is a noteworthy and interesting commentary on the relationship between mathematical ideas and their application in the physical sciences that the study of these strange geometries preceded their application in physics by a very long interval (indeed, by more than five decades).

★ ★ ★

In the problems below, we refer to the Minkowski distance by the symbol d_M. Thus, if $A\,(a,b)$ and $B\,(c,d)$ are two points in Minkowski space, then $d_M\,(A,B) = \sqrt{-(a-c)^2 + (b-d)^2}$.

EXERCISES

3.8.1 With reference to a fixed point O, sketch the locus of points P at zero Minkowski distance from O; that is, the locus

$$\{P : d_M(O,P) = 0\}\,.$$

Also sketch the locus of points at an imaginary Minkowski distance from O.

3.8.2 Let O be a fixed point and r a positive real number. Sketch the Minkowski circle

$$\{P : d_M(O,P) = r\}\,.$$

3.8.3 Let A, B be fixed points in Minkowski space. Sketch the locus

$$\{P : d_M(A,P) = d_M(B,P)\}\,.$$

3.8.4 In this problem, we shall make further changes to the distance formula.

Let $P\,(x,y)$ and $Q\,(x+h, y+k)$ be two points close to one another (this means that $h, k \approx 0$). We take the distance s between P and Q to be given by

$$s = \frac{\sqrt{h^2 + k^2}}{1 + x^2 + y^2}.$$

This is the *metric* of the space: it indicates how distance is related to infinitely small displacements in space.

The factor in the denominator gives rise to curious effects; it implies, for instance, that as we move away from the origin, *distances shrink*. A consequence of this is that infinity is not infinitely far away! Indeed, the use of calculus shows that the distance from $(0,0)$ to $(\infty,0)$, travelling along the x-axis, is

$$\int_0^\infty \frac{dx}{1+x^2} = \left[\tan^{-1} x\right]_0^\infty = \frac{\pi}{2},$$

a finite number. It turns out that this geometry is equivalent to one where the world is a finite disk.

What are the shortest lines in this geometry?

Suggested Reading For further reading on this topic, we list the following titles.

[1] Edwin A Abott. *Flatland* (Dover reprint, New York, 1952).

[2] Arthur Eddington. *Space, Time and Gravitation* (Harper & Row, New York, 1959).

[3] Rudolf v B. Rucker. *Geometry, Relativity and the Fourth Dimension* (Dover, New York, 1977).

3.9 A potpourri of problems

EXERCISES

3.9.1 Let M be the midpoint of the base BC of $\triangle ABC$. It is well-known that triangles ABM and AMC have equal areas. Find a way of dissecting $\triangle ABM$ into as few pieces as possible, so that the pieces can be rearranged to tile $\triangle AMC$ exactly (no gaps or overlaps).

3.9.2 Given an arbitrary $\triangle ABC$, find a way of inscribing an equilateral triangle within the given triangle. That is, find the points D, E and F on the sides BC, CA and AB, respectively, such that $\triangle DEF$ is equilateral. Show that there are, in general, infinitely many such inscribed triangles.

COMMENT. An interesting but difficult problem would be to find the locus of centroids of these equilateral triangles.

3.9.3 Let $\triangle ABC$ be equilateral with the circumcircle Γ, and let D be any point on minor arc BC. Show that $DA = DB + DC$.

3.9.4 Let Γ be an a circle, and let $ABCDEF$ be a convex hexagon inscribed within Γ, such that the diagonals AD, BE and CF meet in a point P. Suppose further that these diagonals are equally inclined to one another, i.e., angles APB, BPC, CPD, DPE, EPF and FDA are each equal to $60°$. Show that

$$PA + PC + PE = PB + PD + PF.$$

3.9.5 In $\triangle ABC$, let the internal bisectors of $\angle A$, $\angle B$ and $\angle C$ meet the opposite sides (BC, CA and AB) in the points D, E and F, respectively. Show that if $\angle B = 120°$, then $\angle DEF = 90°$.

HINT. Use the property $AB/AC = BD/DC$.

3.9.6 Let $\triangle ABC$ be an isosceles, with $\angle A = 36°$, $\angle B = \angle C = 72°$. Let the bisector of $\angle ABC$ meet the side AC at D. Show that AD/DC is the "golden ratio", i.e., the number $(\sqrt{5}+1)/2$. Deduce that

$$\cos 72° = \frac{\sqrt{5}-1}{4}.$$

Use the relation to find a ruler-and-compass construction for a regular pentagon.

3.9.7 Let P be a point within the interior of $\triangle ABC$, and let AP, BP, CP meet the sides BC, CA, AB at points D, E, F respectively. Prove that

$$\frac{AP}{PD} = \frac{AF}{FB} + \frac{AE}{EC}.$$

3.9.8 Let X, Y and Z be points on the sides BC, CA and AB of $\triangle ABC$, such that

$$\frac{BX}{XC} = \frac{CY}{YA} = \frac{AZ}{ZB} = \frac{1}{2}$$

(i.e., X, Y, Z are points of trisection of the respective sides). The lines AX, BY, CZ enclose a triangle within $\triangle ABC$.

Prove that the area of this triangle is 1/7 of the area of $\triangle ABC$.

3.9.9 **Ceva's Theorem** Let P be a point lying within $\triangle ABC$, and let AP, BP, CP meet the sides BC, CA, AB at X, Y and Z, respectively. Prove that

$$\frac{BX}{XC}\cdot\frac{CY}{YA}\cdot\frac{AZ}{ZB}=1.$$

3.9.10 Let AD be an altitude of an acute-angled triangle ABC, and let P be a point on AD. Let BP meet CA at Y, and let CP meet AB at Z. Prove that $\angle ADZ = \angle ADY$.

3.9.11 Let $ABCD$ be an isosceles trapezium, with AB parallel to DC and $AD = BC$, and $\angle ADC = \angle BCD = 36°$. Let R denote the radius of the circumcircle of $ABCD$. Prove that $DC - AB = R$.

3.9.12 Let $ABCD$ be a square of side 1 unit and let D' be a point on BC. Let the square be reflected in the perpendicular bisector of DD', and let the image of A be A'. Let $A'D'$ meet BC at E. Show that the perimeter of $\triangle EBD'$ is 2 units and that $\angle EDD' = 45°$.

3.9.13 Let $\mathcal{P}$ and $\mathcal{P}'$ be polygons with equal area. Show that $\mathcal{P}$ can be dissected into a finite number of pieces such that the pieces can be rearranged so as to cover $\mathcal{P}'$ exactly, with no gaps or overlaps.

3.9.14 The following result is a well-known one: *If $\mathcal{C}$ is a circle with centre O and radius r, and P is a point outside the circle with $OP = d$, then the length t of the tangent from P to $\mathcal{C}$ is given by $t^2 = d^2 - r^2$.* Use the result to show the following.

Let $\mathcal{C}_1$ and $\mathcal{C}_2$ be two circles that intersect in points A and B. Let P be a point such that the tangents from P to $\mathcal{C}_1$ and $\mathcal{C}_2$ have equal length. Prove that the points P, A and B are collinear.

3.9.15 (Continuation) Let $\mathcal{C}_1$, $\mathcal{C}_2$ and $\mathcal{C}_3$ be three mutually intersecting circles. Show that their three common chords pass through a point.

3.9.16 Let BC be a chord of a circle Γ, and let S_1, S_2 be the two arcs of Γ cut off by BC. Let M be the midpoint of S_2, and consider all possible circles, Ω, that touch both S_1 and BC. Show that the length, $t_{M\Omega}$, of the tangent from M to Ω is constant for all such Ω.

3.9.17 (Continuation) Let $\mathcal{C}$ be a circle with diameter AB, and let P be a point on AB. Draw PQ perpendicular to AB with Q on $\mathcal{C}$. Let $\mathcal{C}_1$ be the circle that touches PQ, touches PA at S and touches $\mathcal{C}$ internally (on arc AQ); and let $\mathcal{C}_2$ be the circle that touches PQ, touches PB at T and touches C internally (on arc BQ). Prove that $\angle SQT = 45°$.

3.9.18 Let AB and CD be non-intersecting chords of a circle $\mathcal{C}$, and let $\mathcal{A}$ denote the arc AB not containing C and D. For any arbitrary point $P \in \mathcal{A}$, let X and Y be the points where PC and PD meet AB. Let $AX = u$, $XY = v$, $YB = w$. Show that uw/v has the same value for all positions of P.

3.9.19 (Continuation) Let AB be a chord of a circle $\mathcal{C}$ and let M be the midpoint of AB. Let CD and EF be two other chords of C that pass through M, and let DE and CF meet AB at points X and Y, respectively. Show that $MX = MY$. (This result is known as the **Butterfly Theorem.**)

3.9.20 Let $\triangle ABC$ be acute-angled with circumcentre O, and let BO and CO meet AC and AB at D and E, respectively. Suppose that $\angle BAC = \angle AED = \angle DEO$. Find the angles of $\triangle ABC$.

3.9.21 In the isosceles $\triangle ABC$, we have $\angle A = 100°$ and $\angle B = \angle C = 40°$. Side AB is extended to D so that $AD = BC$. Find the size of $\angle BCD$.

3.9.22 In the isosceles $\triangle ABC$, we have $\angle A = 20°$ and $\angle B = \angle C = 80°$. Points D and E lie on sides AC and AB, respectively, such that $\angle DBC = 60°$ and $ECB = 50°$. Find the size of $\angle BDE$.

3.9.23 **The Pizza Theorem** Through an arbitrary point P in a circle $\mathcal{C}$, four chords are drawn so that each of the eight angles formed at P is 45°. The regions into which $\mathcal{C}$ is divided are now numbered 1, 2, 3, ..., 7, 8 in cyclic order. Show that the total area of the regions marked 1, 3, 5, 7 is equal to the total area of the regions marked 2, 4, 6, 8.

3.9.24 **Centroid of a triangle** Let $\triangle ABC$ be given. Show that the following two configurations possess the same centre of gravity:

(a) three particles of equal weight at the three vertices of the triangle, held together by a thin wire whose weight is negligible;
(b) a triangular lamina of uniform thickness whose outline coincides with the boundary of $\triangle ABC$.

3.9.25 **Centroid of a quadrilateral** What happens in the case of a quadrilateral $ABCD$? Show that the centres of gravity of the following two configurations:

(a) four particles of equal weight placed at the vertices of the quadrilateral and held together by a thin wire of uniform width,
(b) a lamina of uniform thickness whose outline coincides with that of the quadrilateral,

are, in general, *not* the same. Specifically, show that the centres of gravity in cases (a) and (b) can be found as follows:

(a) Let the midpoints of AB, BC, CD and DA be P, Q, R and S respectively. Then, $PQRS$ is a parallelogram and the required centre of gravity lies at the point of intersection of the diagonals of the parallelogram.

(b) Let the points of trisection of the four sides be as follows: P and Q for side AB; R and S for side BC; S and T for side CD; U and V for side DA, the labelling being such that the points lie in the given order: A, P, Q, B on AB, and so on. Let the lines QR, ST, UV and WP be drawn. These lines, when extended, enclose a parallelogram. Then, the centre of gravity of the quadrilateral $ABCD$ lies at the point of intersection of the diagonals of the parallelogram.

3.10 References

[1] Coxeter, H S M and Greitzer, S L. *Geometry Revisited*, MAA.

[2] Durell, C V. *Modern Pure Geometry*, Macmillan.

[3] Johnson, R A. *Advanced Euclidean Geometry*, Dover.

[4] Maxwell, E A. *Geometry For Advanced Pupils*.

[5] Pedoe, D. *Circles: A Mathematical View*, Dover.

[6] Tuller, A. *A Modern Introduction to Geometries*, Van Nostrand.

References [1], [2], [3] are highly recommended—they contain an astonishing wealth of material, particularly [3]. Reference [6] has a good description of many non-Euclidean geometries, and [5] has material on inversion and constructions which use only compasses.

Chapter 4

Faster, Higher, Stronger!

4.1 Preliminaries

We have already come across inequalities in Chapter 2 during our search for approximations to various irrational numbers. In this chapter, we shall study inequalities in a more systematic manner. First, we present some appetizers.

Problems

Which is greater:

(a) $\dfrac{2000}{2001}$ or $\dfrac{2001}{2002}$?

(b) 11^{33} or 2001^{10}?

(c) $2.000000001 + \dfrac{1}{2.000000001}$ or $2.000000002 + \dfrac{1}{2.000000002}$?

Solutions

(a) Clearly, 2001/2002 is the greater fraction; for

$$2001^2 > 2001^2 - 1 = 2000 \times 2002, \quad \therefore \quad \frac{2001}{2002} > \frac{2001}{2002}.$$

(b) We have,

$$11^{33} = (11^3)^{11} > (11 \times 10^2)^{11} > (2^{10})^{11} = (2^{11})^{10} > (2001)^{10}.$$

If a more convincing proof is called for, then using Derive, we have

$$\begin{aligned} 11^{33} &= 2322515441988780814100176779630913 1 \\ 2001^{10} &= 1029131535373448067360960180020001. \end{aligned}$$

The larger number is roughly 20 times the smaller one.

(c) Write, α for 2.000000001 and β for 2.000000002; then $\beta > \alpha$, and we need to check which is greater: $\alpha + 1/\alpha$ or $\beta + 1/\beta$. We have

$$\begin{aligned} \left(\beta + \frac{1}{\beta}\right) - \left(\alpha + \frac{1}{\alpha}\right) &= (\beta - \alpha) + \left(\frac{1}{\beta} - \frac{1}{\alpha}\right) \\ &= (\beta - \alpha) + \left(\frac{\alpha - \beta}{\alpha\beta}\right) \\ &= (\beta - \alpha)\left(1 - \frac{1}{\alpha\beta}\right) \\ &= \frac{(\beta - \alpha)(\alpha\beta - 1)}{\alpha\beta}. \end{aligned}$$

Since $\beta > \alpha > 1$, we have $\alpha\beta > 1$; $\therefore$ the last quantity is positive, implying that $\beta + 1/\beta$ exceeds $\alpha + 1/\alpha$.

4.2 Basic rules

We start by listing the basic rules that underlie the theory of inequalities. The rules are not independent of one another, and they must not be thought of as axioms. In the discussion below, **R** refers throughout to the set of real numbers.

1. A relation '>' ("greater than") may be defined for **R**, such that for any two distinct real numbers a and b, either $a > b$ or $b > a$.
2. The relation '>' is *transitive*; in other words, $a > b$ together with $b > c$ implies that $a > c$.
3. The set of real numbers **R** can be partitioned as

$$\mathbf{R} = \text{Pos} \cup \{0\} \cup \text{Neg}$$

(respectively: positive numbers, zero, negative numbers), such that for every $a \in$ Pos and every $b \in$ Neg, we have $a > 0 > b$.

4. (a) If $a, b \in \mathsf{Pos}$, then $a+b \in \mathsf{Pos}$ and $ab \in \mathsf{Pos}$;
   (b) if $a, b \in \mathsf{Neg}$, then $a+b \in \mathsf{Neg}$ and $ab \in \mathsf{Pos}$;
   (c) if $a \in \mathsf{Pos}, b \in \mathsf{Neg}$ then $ab \in \mathsf{Neg}$.
5. If $a > b$ and $c \in \mathbf{R}$, then $a+c > b+c$.
6. If $a > b$ and $c \in \mathsf{Pos}$, then $ac > bc$; but if $c \in \mathsf{Neg}$, then $ac < bc$.
7. If $a, b \in \mathsf{Pos}$ and $a > b$, then $1/a < 1/b$.

⋆ ⋆ ⋆

The proofs of a great many inequalities[1] ultimately depend upon the following seemingly trivial observation:

> (∗) *The square of any real number x is non-negative; that is, $x^2 \geq 0$ for all $x \in \mathbf{R}$.*

For instance, consider the following rather obvious statement: *For all real numbers x we have $(x-1)^2 \geq 0$, with equality iff $x = 1$.* (Here "iff" as a short-form for 'if and only if'.) We derive below a significant consequence from this statement.

We have, $(x-1)^2 \geq 0$, $\quad\therefore\; x^2 - 2x + 1 \geq 0$, $\quad\therefore\; x^2 + 1 \geq 2x$. Suppose that $x > 0$. We may now divide by x without losing the direction of the inequality; we obtain the following:

$$x + \frac{1}{x} \geq 2 \quad \text{for all } x > 0, \text{ with equality iff } x = 1.$$

EXERCISE Show that if $x < 0$, then $x + 1/x \leq -2$, with equality iff $x = -1$.

A geometric proof Here is a pretty geometric proof of the above result. For a given positive number x, let A, B and C be points on a line ℓ, in that order, with $AB = x$ and $BC = 1/x$ (see Figure 4.1); then $AC = x + 1/x$.

Let Γ be the circle on AC as the diameter; let the line through B, perpendicular to ℓ, meet $\mathcal{C}$ at D and E; then $DB = BE$, by symmetry. By the chord theorem, $DB \cdot BE = AB \cdot BC = x \cdot 1/x = 1$, $\therefore\; DB = 1 = BE$, $\;\therefore\; DE = 2$.

Since DE is a chord of $\mathcal{C}$ and AC is a diameter, we get $AC \geq DE$, implying that $x + 1/x \geq 2$. □

[1] Perhaps all of them!

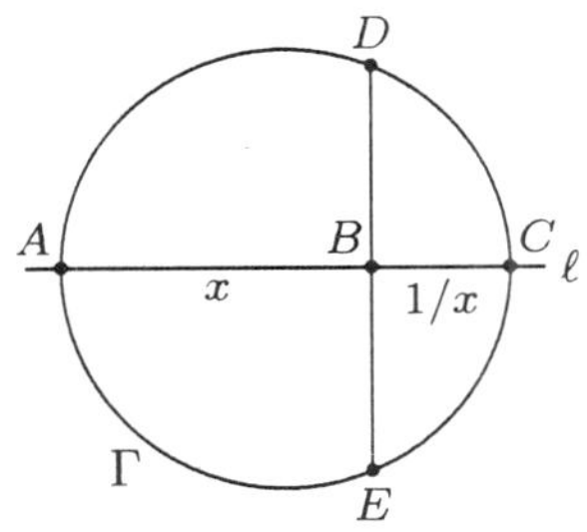

Figure 4.1. *Geometric proof of an inequality*

Some corollaries Several nice corollaries may be deduced from the above result, for instance: *Let* a, b, c *be positive real numbers; then*

$$(a+b+c)\left(\frac{1}{a}+\frac{1}{b}+\frac{1}{c}\right) \geq 9,$$

with equality iff $a = b = c$. For proof, we simply expand the expression on the left. We obtain

$$3+\left(\frac{a}{b}+\frac{b}{a}\right)+\left(\frac{b}{c}+\frac{c}{b}\right)+\left(\frac{c}{a}+\frac{a}{c}\right).$$

Each bracketed term is of the form $x + 1/x$ for some $x > 0$, therefore not less than 2. Thus, the quantity on the left is not less than 9. □

EXERCISE

4.2.1 Show that if $a, b, c, \ldots$ are n positive real numbers, then

$$(a+b+c+\cdots)\left(\frac{1}{a}+\frac{1}{b}+\frac{1}{c}+\cdots\right) \geq n^2,$$

with equality iff $a, b, c, \ldots$ are all equal to one another.

Another application: the AM–GM inequality Here is another application of the basic fact (∗). Let a, b be positive numbers. Then $(\sqrt{a}-\sqrt{b})^2 \geq 0$, with equality iff $a = b$. This yields, on simplification, $a + b - 2\sqrt{ab} \geq 0$, with equality iff $a = b$. Therefore, transposing terms,

$$\frac{a+b}{2} \geq \sqrt{ab}, \text{ with equality iff } a = b.$$

More can be said, by using the non-negativity of $(1/\sqrt{a} - 1/\sqrt{b})^2$. We obtain: $1/a+1/b-2/\sqrt{ab} \geq 0$, with equality iff $a = b$. Therefore, transposing terms,

$$\sqrt{ab} \geq \frac{2}{1/a + 1/b}.$$

These results may be stated in words using the terminology of means. Given n positive real numbers a, b, c, ..., their arithmetic mean (AM), geometric mean (GM) and harmonic mean (HM) are defined thus:

$$\text{AM} = \frac{a+b+c+\cdots}{n},$$

$$\text{GM} = \sqrt[n]{a\,b\,c\cdots},$$

$$\text{HM} = \frac{n}{1/a + 1/b + 1/c + \cdots}.$$

Observe that the HM is the reciprocal of the AM of the reciprocals of a, b, c, ...

Using this terminology, we have the following: *For any two positive real numbers a and b,*

$$\text{AM} \geq \text{GM} \geq \text{HM},$$

with equality iff $a = b$. This is known as the AM–GM–HM inequality; or, if we leave out the HM, as the AM–GM inequality. The inequality is very widely used in mathematical analysis. It can be proved using geometric ideas, just as we did for the inequality $x+1/x \geq 2$ for $x > 0$; but we leave the details to the reader.

Here is an elementary application of the AM–GM inequality: *If n is an integer greater than 1, then*

$$(2n-1)!^2 < n^{2n-1}.$$

For proof, note that $i \cdot (2n-i) < n^2$ for $i \in \{1, 2, \ldots, n-2, n-1\}$; this comes from a direct application of the AM–GM inequality. For $i = n$, equality holds. Multiplying together the corresponding sides of the n relations, we obtain the required result.

★ ★ ★

The AM–GM inequality holds more generally for any list of non-negative numbers: *If n is a positive integer and $a_1, a_2, \ldots, a_n$ are non-negative real numbers, then*

$$\frac{a_1 + a_2 + \cdots a_n}{n} \geq (a_1 a_2 \ldots a_n)^{1/n},$$

with equality iff the a_is are all equal to one another.

Proof of the AM–GM inequality The proof we give here is different from that given in most texts. Let A and G denote, respectively, the AM and GM of the n numbers $a_1, a_2, \ldots, a_n$. We shall show that $A \geq G$.

We shall assume that the a_is are indexed so that they occur in increasing order:

$$a_1 \leq a_2 \leq \cdots \leq a_n,$$

then $a_1 \leq A \leq a_n$. If $a_1 = A$ or $a_n = A$, then the a_is are all equal, in which case $A = G$ and there is nothing to prove; so we assume that $a_1 < A < a_n$. This implies that the quantity $X = a_1 + a_n - A$ is positive.

We now replace a_1 and a_n by A and X, respectively. The n numbers we have as a result are:

$$A,\ a_2,\ a_3,\ \ldots,\ a_{n-1},\ X.$$

The AM of these numbers is the same as that of the numbers in the earlier collection, because $A + X = a_1 + a_n$. However, their GM is strictly greater than the earlier value, because A and X lie strictly between a_1 and a_n, implying that $AX > a_1 a_n$. (This comes from the following assertion: *If a, b, c, d are positive numbers with $a < b, c < d$ and $a + d = b + c$, then $ad < bc$.* For example, $2 + 10 = 3 + 9$ and $2 \cdot 10 < 3 \cdot 9$. This, in turn, comes from the relations

$$ad = \left(\frac{d+a}{2}\right)^2 - \left(\frac{d-a}{2}\right)^2, \quad bc = \left(\frac{c+b}{2}\right)^2 - \left(\frac{c-b}{2}\right)^2.$$

Since $d - a > c - b$, it follows that $ad < bc$. This argument, with a, b, c, d replaced by a_1, A, X, a_n respectively, yields the result $AX > a_1 a_n$.)

The replacement therefore preserves the AM but results in a strictly increased value for the GM.

We now iterate the procedure: at each stage, we replace the least and greatest numbers in the updated collection by the AM of the n numbers and the value of (least number plus greatest number minus AM), and we continue doing this as long as the numbers in the collection are not all equal. Each iteration yields a collection of numbers in which the number of entries equal to A has gone up by at least 1. Therefore, after $n - 1$ or fewer

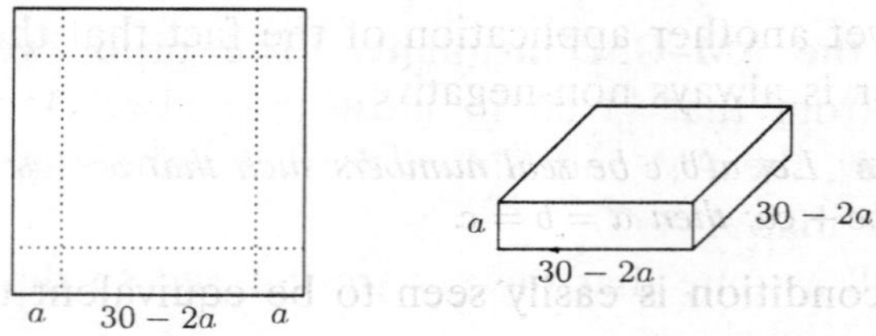

Figure 4.2. *Maximizing the volume of an open box*

iterations, we reach a stage when the entries are all equal to A. Let G_i represent the GM at the ith stage, so that $G_0 = G$. Then,

$$A = G_{n-1} \geq G_{n-2} \geq \cdots \geq G_1 \geq G_0 = G,$$

so $A \geq G$, as required. The inequality is strict unless the a_is are all equal. □

★ ★ ★

The AM–GM inequality is very powerful, and, when used with some skill, it can yield very elegant results. Here is one such result, often posed as a problem in early courses on calculus.

> **Problem**
> *From each corner of a square measuring $30' \times 30'$ a square measuring $a' \times a'$ is removed. The sides are now folded and a box with dimensions $(30-2a)' \times (30-2a)' \times a'$ is obtained. For which a is the volume of the box the largest possible?* [See Figure 4.2.]

Solution The volume $V(a)$ is given by $V(a) = a(30-2a)^2$. Note that $0 \leq a \leq 15$; over this domain, we need to maximize V.

Consider the three non-negative numbers: $4a$, $30-2a$, and $30-2a$ whose AM is $60/3$ or 20. The AM–GM inequality implies that their GM cannot exceed 20:

$$\sqrt[3]{4a\,(30-2a)\,(30-2a)} \leq 20 \quad \text{for } 0 \leq a \leq 15,$$

with equality iff $4a = 30 - 2a$, i.e., iff $a = 5$. It follows that for $0 \leq a \leq 15$ the following inequality holds:

$$4a(30-2a)^2 \leq 20^3,$$

or $V(a) \leq 2000$, with equality iff $a = 5$. The optimal dimensions of the box are therefore $20' \times 20' \times 5'$, and the maximum possible volume is $V_{\max} = 2000$. □

★ ★ ★

Here is yet another application of the fact that the square of a real number is always non-negative.

> **Claim** *Let a, b, c be real numbers such that $a^2 + b^2 + c^2 = ab + bc + ca$; then $a = b = c$.*

The given condition is easily seen to be equivalent to

$$(a-b)^2 + (b-c)^2 + (c-a)^2 = 0.$$

Since a sum of non-negative numbers is 0 iff each number is 0, the condition holds iff $0 = a - b = b - c = c - a$, i.e., iff $a = b = c$. □

⋆ ⋆ ⋆

More generally, we have the following: *For any n real numbers a_1, a_2, a_3, ..., a_{n-1} and a_n, we have*

$$a_1^2 + a_2^2 + a_3^2 \cdots + a_n^2 \geq a_1a_2 + a_2a_3 + \cdots + a_{n-1}a_n + a_na_1$$

(with n summands on both sides), and equality holds iff $a_1 = a_2 = \cdots = a_n$. The proof runs along the same lines.

Non-negativity of quadratic polynomials Another elementary propo sition often used in proving inequalities is the following:

> **Proposition** *Let P be a quadratic polynomial with real coefficients, say $P(x) = ax^2 + bx + c$ where $a \neq 0$. Suppose that $P(x)$ never changes sign: for $x \in \mathbf{R}$ it is either always non-negative or always non-positive; then $b^2 - 4ac \leq 0$. Moreover, if $P(x)$ is never 0 (∴ either strictly positive or strictly negative for all $x \in \mathbf{R}$), then $b^2 - 4ac < 0$.*

The proof follows from the 'completing-the-square' identity. We have

$$4a\,P(x) = 4a^2x^2 + 4abx + 4ac = (2ax + b)^2 + (4ac - b^2).$$

From this, it follows that $4a\,P(x) \geq 4ac - b^2$ for all x and so:

(a) if $a > 0$, then $P(x) \geq c - b^2/4a$ for all x; and

(b) if $a < 0$, then $P(x) \leq c - b^2/4a$ for all x.

In both cases, the equality sign holds iff $x = -b/2a$.

So if $P(x) > 0$ for all x, then it must be the case that $a > 0$ and $c - b^2/4a > 0$; and if $P(x) < 0$ for all x, then it must be the case that $a < 0$ and $c - b^2/4a < 0$. In both cases we obtain, by multiplication by a, the condition $b^2 - 4ac < 0$.

The Cauchy–Schwartz inequality Here is an impressive example of how the above proposition is put to use: in proving the famous Cauchy–Schwartz inequality.

> **Cauchy–Schwartz Inequality** *Let n be a positive integer, and let $a_1, a_2, \ldots, a_n$ and $b_1, b_2, \ldots, b_n$ be two lists of real numbers. Let*
>
> $$\begin{aligned} A &= a_1^2 + a_2^2 + \cdots + a_n^2, \\ B &= a_1 b_1 + a_2 b_2 + \cdots + a_n b_n, \\ C &= b_1^2 + b_2^2 + \cdots + b_n^2. \end{aligned}$$
>
> *Then $AC \geq B^2$, with equality iff either for some real number t we have $a_i = tb_i$ for all i, <u>or</u> $b_i = 0$ for all i.*

The proof is surprisingly easy. If $a_i = 0$ for all i, then we get $A = 0$ and $B = 0$, so the inequality $B^2 - AC \leq 0$ holds trivially. If $b_i = 0$ for all i, then we get $B = 0$ and $C = 0$, and once again the inequality $B^2 - AC \leq 0$ holds, trivially. Now suppose that the a_is are not all 0, and the b_is are not all 0; then $A > 0$. Consider the quadratic function $P(x)$ defined for real numbers x thus:

$$P(x) = (a_1 x + b_1)^2 + (a_2 x + b_2)^2 + \cdots + (a_n x + b_n)^2 .$$

Since $P(x)$ is a sum of squares, we have $P(x) \geq 0$ for all x. On opening the brackets, we find that $P(x) = Ax^2 + 2Bx + C$, with A, B and C as given above. Since $A > 0$, the proposition proved above can be invoked for the quadratic polynomial $Ax^2 + 2Bx + C$, and we immediately obtain the result $B^2 - AC \leq 0$. This is precisely the Cauchy–Schwartz inequality.

Under what circumstances will equality hold (i.e., $B^2 - AC = 0$)? If equality holds, then clearly $P(x)$ is a constant times the square of a linear expression and, therefore, can assume the value 0. But from the form of $P(x)$ it is readily seen that $P(x)$ can assume a value of 0 iff there exists a value of x for which $a_i x + b_i = 0$ for all i; equivalently, if there exists a value of t for which $a_i = tb_i$ for all i.

Conversely, if such a relation exists between the a_i and b_i, then $P(x)$ is a constant times the square of a linear expression and in this case $B^2 - AC = 0$. □

Two applications of Cauchy–Schwartz The following proposition, mentioned earlier and proved via the result that $x + 1/x \geq 2$ for

$x > 0$, is an immediate corollary of the Cauchy–Schwartz inequality: *Let $a, b, c, \ldots$ be any n positive numbers, then*

$$(a+b+c+\cdots)\left(\frac{1}{a}+\frac{1}{b}+\frac{1}{c}+\cdots\right) \geq n^2,$$

with equality iff $a = b = c = \cdots$. The proof follows via the substitutions $a_1 = \sqrt{a}$, $a_2 = \sqrt{b}$, $a_3 = \sqrt{c}$, ... and $b_1 = 1/a_1$, $b_2 = 1/a_2$, $b_3 = 1/a_3$ ... in the statement of the theorem.

A more difficult result, which however we do not prove here, is the following.

> **Proposition** *Let $P(x)$ be a polynomial with real coefficients such that $P(x) \geq 0$ for $x \in \mathbf{R}$. Then there exist polynomials $f_1(x)$, $f_2(x)$, ..., all with real coefficients, such that $P(x) = f_1(x)^2 + f_2(x)^2 + \cdots$.*

4.3 Problems

4.3.1 Show that amongst all rectangles with perimeter $4a$, the square with side a has the largest area.

4.3.2 Show that amongst all rectangles with area a^2, the square with side a has the least perimeter.

4.3.3 Show that amongst all triangles with perimeter $3p$, the equilateral triangle, with side p, has the largest area.

4.3.4 (Continuation) Show that the following inequality holds between the perimeter p and the area Δ of a triangle:

$$p^2 \geq 12\sqrt{3}\,\Delta.$$

4.3.5 Let $x_1, x_2, \ldots, x_n$ be non-negative real numbers satisfying the condition $x_1 + x_2 + \cdots + x_n = k$ for some given positive number k. Determine the maximum and minimum possible values of $x_1^2 + x_2^2 + \cdots + x_n^2$.

4.3.6 Chebychef's inequality Let $a_1, a_2, \ldots, a_n$ and $b_1, b_2, \ldots, b_n$ be two lists of n real numbers each, both arranged in decreasing order:

$$a_1 \geq a_2 \geq \cdots \geq a_n, \qquad b_1 \geq b_2 \geq \cdots \geq b_n.$$

Let $\overline{\mathsf{A}}$, $\overline{\mathsf{B}}$ and $\overline{\mathsf{AB}}$ be the AMs of the collections $\{a_i\}$, $\{b_i\}$ and $\{a_ib_i\}$, respectively:

$$\begin{aligned}\overline{\mathsf{A}} &= (a_1 + a_2 + \cdots)/n,\\ \overline{\mathsf{B}} &= (b_1 + b_2 + \cdots)/n,\\ \overline{\mathsf{AB}} &= (a_1b_1 + a_2b_2 + \cdots)/n.\end{aligned}$$

Show that $\overline{\mathsf{AB}} \geq \overline{\mathsf{A}} \cdot \overline{\mathsf{B}}$. In other words, *the mean of the products is greater than or equal to the product of the means.* If the a_is are in decreasing order while the b_is are in increasing order, or *vice versa*, then the inequality is reversed: $\overline{\mathsf{AB}} \leq \overline{\mathsf{A}} \cdot \overline{\mathsf{B}}$.

4.3.7 Let a and b be positive numbers. Find proofs via geometry for the following string of inequalities:

$$\sqrt{\frac{a^2+b^2}{2}} \geq \frac{a+b}{2} \geq \sqrt{ab} \geq \frac{2ab}{a+b}.$$

4.3.8 Show that for $0 \leq x \leq 1$, the function $f(x) = x^2 - x^3$ takes its largest value when $x = 2/3$.

4.3.9 It is known that the two roots of the equation $ax^2+bx+c=0$ are real iff $b^2 - 4ac \geq 0$. Find the condition for all three roots of the cubic equation $ax^3 + bx^2 + c = 0$ to be real. (Here a, b, c are real numbers. Note that the equation lacks the 'x'-term.)

4.3.10 Let $\mathcal{B}$ be an open box with a square base and with a total surface area A. Let V be the volume of $\mathcal{B}$. Show that $108\,V^2 \leq A^3$.

4.3.11 A collection of positive integers has a sum of 1999. What is their largest possible product?

4.3.12 (Continuation) For a given positive integer $n \geq 4$, let $f(n)$ denote the largest possible product of a collection of positive integers whose sum is n. Find an algorithm for computing $f(n)$.

4.3.13 Let $a, b, c, \ldots$ be n given quantities. Show that the value of x, which minimizes the quadratic function

$$Q(x) = (a - x)^2 + (b - x)^2 + (c - x)^2 + \cdots,$$

is $(a + b + c + \cdots)/n$, i.e., the AM of $a, b, c, \ldots$.

4.3.14 Let $a, b, c, \ldots$ be n given quantities. Show that the value of x which minimizes the function

$$Q(x) = |a - x| + |b - x| + |c - x| + \cdots$$

is the *median* of $a, b, c, \ldots$. (The median of a list of n numbers $a_1, a_2, \ldots, a_n$ is defined to be any number M such that the cardinalities of the two sets

$$\{i : a_i \leq M\}, \qquad \{i : a_i \geq M\},$$

are equal; that is, it is the value which would "lie in the middle" if the numbers were placed in ascending order.

EXAMPLE. The median of the list 2, 3, 5 is 3; the median of the list 1, 2, 4, 7, 9 is 4; and the median of the list 1, 2, 5, 9 is any number M lying between 2 and 5.)

4.3.15 Let $P(a, b)$ be a point in the first quadrant of the coordinate plane $(a, b > 0)$; let O denote the origin. Let ℓ be a line through P with a negative slope meeting the axes at A and B, respectively. Show that $\text{Area}(\triangle AOB) \geq 2ab$.

4.3.16 (Continuation) Let $\mathcal{R}$ be a rectangle inscribed in a triangle $\mathcal{T}$, with two vertices on one side of the triangle and one vertex each on the other two sides. Show that $\text{Area}(\mathcal{R}) \leq \frac{1}{2}\,\text{Area}(\mathcal{T})$.

4.3.17 Show that the largest possible area of a rectangle inscribed in a circle of radius r is $2r^2$.

4.3.18 (Continuation) For fixed numbers $a, b > 0$, consider all possible rectangles $\mathcal{R}$ inscribed in the ellipse with the equation

$$\frac{x^2}{a^2} + \frac{y^2}{b^2} = 1,$$

the sides of the rectangle being parallel to the axes of the ellipse. Show that $\text{Area}(\mathcal{R}) \leq 2ab$.

4.3.19 Show that the volume of a right circular cylinder inscribed in a sphere is less than or equal to $1/\sqrt{3}$ of the volume of the sphere.

4.3.20 Show that the largest possible volume of a right circular cone inscribed in a sphere is $8/27$ of the volume of the sphere.

4.3.21 Show that the following inequality holds for all integers $n > 1$:

$$\frac{1}{2} < \frac{1}{n+1} + \frac{1}{n+2} + \cdots + \frac{1}{2n-1} + \frac{1}{2n} < 1.$$

4.3.22 Show that the following inequality holds for all integers $n \geq 4$:

$$1 < \frac{1}{n+1} + \frac{1}{n+2} + \cdots + \frac{1}{3n-1} + \frac{1}{3n} < 2.$$

4.3.23 Let a, b, c be positive numbers such that $abc = 1$. Show that

$$\frac{1}{a^3(b+c)} + \frac{1}{b^3(c+a)} + \frac{1}{c^3(a+b)} \geq \frac{1/a + 1/b + 1/c}{2}.$$

(This is a generalization of Problem 2 from the IMO held in Canada, in July 1995.)

4.3.24 Let $0 \leq a, b, c, d \leq 1$. Find the largest possible value of the sum

$$\frac{a}{bcd+1} + \frac{b}{cda+1} + \frac{c}{dab+1} + \frac{d}{abc+1}.$$

4.3.25 Show that for any real number $x > 0$ and any rational number $t > 1$, we have:

(a) $(1+x)^t > 1 + tx$.

(b) $(1+x)^t < \dfrac{1}{1-tx}$, provided that $tx < 1$.

4.3.26 [*] Let a, b, c, d be positive numbers. Let X be the AM of the six numbers: ab, ac, ad, bc, bd, cd; and let Y be the AM of the four numbers: abc, abd, acd, bcd. Prove that $\sqrt{X} > \sqrt[3]{Y}$.

4.3.27 For positive integers n, let u_n be defined as:

$$u_1 = 1, \quad u_{n+1} = u_n + \frac{1}{u_n} \text{ for } n \geq 1.$$

Thus, $u_2 = 2$, $u_3 = 2.5$, $u_{10} = 4.57$, ...

Finding an exact formula for u_n is beyond the scope of the book, so we consider a simpler problem: that of *finding a good approximation for* u_n. Show that:

- $u_n \approx v_n$, where $v_n = \sqrt{2n}$;
- $u_n \approx w_n$, where $w_n = \sqrt{2n} + \sqrt{2/n}$.

(The second approximation is much superior to the first.)

For comparison, we give the values of u_n and w_n for $n = 20$, 40, ..., 100.

n	20	40	60	80	100
u_n	6.42	9.03	11.04	12.72	14.21
w_n	6.64	9.18	11.14	12.81	14.28

Additional data: for $n = 10000$, $u_n = 100.01991$ and $w_n = 100.02$.

4.3.28 Prove that for $k > 0$ we have the inequality

$$\frac{1}{2\sqrt{k+1}} < \sqrt{k+1} - \sqrt{k} < \frac{1}{2\sqrt{k}}.$$

Use this to show that if S_n is given by

$$S_n = 1 + \frac{1}{\sqrt{2}} + \cdots + \frac{1}{\sqrt{n}},$$

then

$$2\left(\sqrt{n+1} - 1\right) < S_n < 2\sqrt{n} - 1.$$

EXAMPLE. For $n = 100$ this gives $18.1 < S_{100} < 19$. In fact, $S_{100} \approx 18.59$.

It follows that a reasonably good approximation for S_n is the arithmetic mean of the two bounds given above; that is,

$$S_n \approx \sqrt{n} + \sqrt{n+1} - 1.5.$$

(For $n = 100$ this gives $S_{100} \approx 18.55$, which is close to the true value.)

★ ★ ★

The last problem concerns an extremely interesting question: that of finding good approximations ('asymptotic formulae') for the partial sums of a given series. Empirically, it turns out that a very good approximation for S_n is given by the formula $T_n = \sqrt{n} + \sqrt{n+1} - 1.46$. The following data, obtained using a computer, shows how close S_n and T_n are.

n	S_n	T_n
1000	61.80099	61.80136
2000	87.99363	87.9939
3000	108.0933	108.0936
4000	125.0388	125.039
5000	139.9682	139.9684

It thus appears that as n tends to ∞, the quantity e_n defined by $e_n = \sqrt{n}+\sqrt{n+1}-S_n$ tends to a limit whose value is approximately 1.46. Extended calculations give the value of the limit as approximately 1.460354. Therefore, we have the following curious result:

$$S_n \approx \sqrt{n} + \sqrt{n+1} - 1.460354.$$

The two sides differ by less than 10^{-6} for $n = 10000$.

4.4 Geometric inequalities

We now take up the theme of inequalities in geometry. The basic facts used are few in number, and all well-known:

(a) The straight line segment connecting two given points is the shortest path between the two points.

(b) The perpendicular from a point P to a line ℓ yields the least distance from P to ℓ.

(c) *Triangle Inequality.* Given any three points A, B, C, we have $AB + BC \geq AC$, with equality iff B lies on the line segment AC.

(d) The longest chord in a circle is its diameter.

(e) Given a smooth curve C and a point P not on the curve, if Q is the point on C which is closest to P, then the tangent to the curve at Q is perpendicular to line QP. (This may be deduced from (b).)

Facts from algebra are also freely used; e.g., the fact that $x+1/x \geq 2$ when $x > 0$.

From these simple beginnings, an impressive collection of results may be deduced. For instance, we can show that the shortest lines on the surface of a sphere are the arcs of great circles (i.e., circles such as the equator and lines of constant longitude). We shall not do so here, however (the reasoning required is rather sophisticated).

4.5 Three examples

We shall first consider three simple problems relating to optimization in a geometric setting. The methods used to solve them are typical of the general approach used.

1. *Given a line ℓ and points A and B on one side of ℓ, show how to find a point P on ℓ such that the length of the path A-P-B (i.e., $AP + PB$) is minimized.* [See Figure 4.3.]
2. *Given a straight river ℓ with parallel banks, and points A and B on opposite sides of the river, find a path A-P-Q-B from A to B with the least total length, where P and Q are points on opposite banks of ℓ and directly opposite one another.* [See Figure 4.5.]
3. *Given an arbitrary triangle ABC in which no angle exceeds 120°, show how to find a point P such that the distance sum $PA + PB + PC$ is minimized.*

Solution to Problem 1. This problem is best solved by using reflection. Let B' denote the image of B under reflection in ℓ. Then we have $AP + PB' = AP + PB$ for every point $P \in \ell$, so our task is to minimize $AP + PB'$.

Observe that $AP + PB' \geq AB'$, with equality iff the points A, P, B' are collinear; therefore, $AP + PB$ is least when A, P, B' are collinear. This indicates how the optimizing point P_0 is found (see Figure 4.3).

Next, observe that AP_0 and BP_0 make equal angles with ℓ; i.e., the path A-P_0-B follows the law of reflection of light (*"angle of incidence equals angle of reflection"*). Stated otherwise, ℓ is the external bisector of angle AP_0B.

The following corollary may be deduced from the above result. *Let ABC be any triangle and let ℓ be the external bisector of $\angle C$. Let*

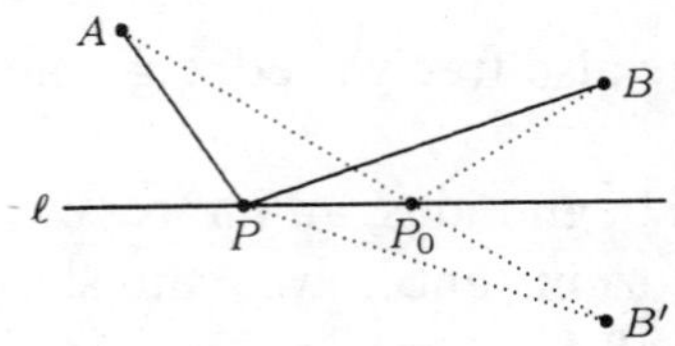

Figure 4.3. *To minimize $AP + PB$.*

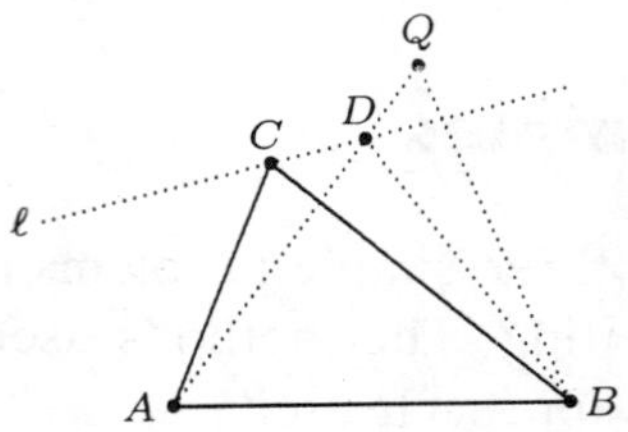

Figure 4.4. *A useful corollary*

Q be any point on the "far side" of ℓ (i.e., Q and A lie on opposite sides of ℓ); then $AQ + QB > AC + BC$. (See Figure 4.4.)

PROOF. Since A and Q lie on opposite sides of ℓ, segment AQ intersects ℓ at some point D. The triangle inequality yields $DB < DQ + QB$. Also, $AC + CB \leq AD + DB$; it follows that

$$AC + CB < AD + DQ + QB$$
$$\therefore \quad AC + CB < AQ + QB.$$

Solution to Problem 2 Let w be the width of the river (see Figure 4.5), and let B' be the image of B under a displacement by w, towards the river and perpendicular to it. (No harm will happen if B' falls within the river!)

Observe that $PQBB'$ is a parallelogram, $\therefore QB = PB'$, $\therefore AP + PQ + QB$ is the same as $AP + w + PB'$. Since w is a constant, the latter quantity is minimized when $AP + PB'$ is minimized, which happens when A, P, B' are collinear. This shows how the optimizing points P_0, Q_0 are to be found.

Solution to Problem 3 The restriction that no angle of the triangle must exceed 120° will be explained later. We only remark here

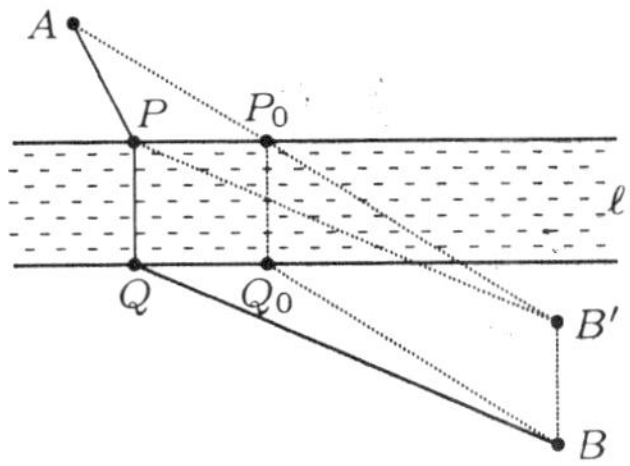

Figure 4.5. *River crossing*

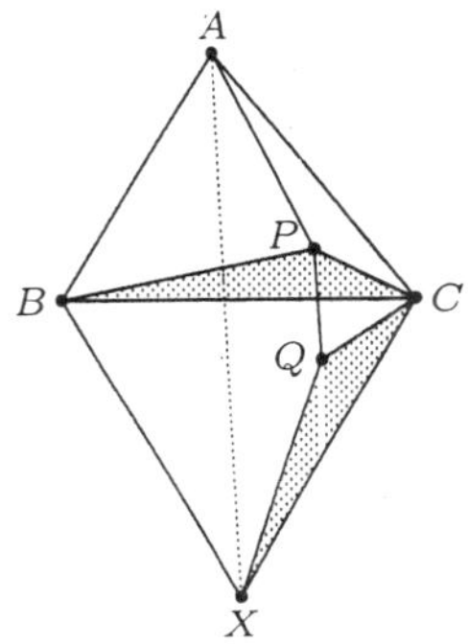

Figure 4.6. $AP + BP + CP \geq AX$

that if, say, $\angle C$ exceeds 120°, then vertex C itself is the minimizing point; that is, $CB + CA \leq PA + PB + PC$ for all points P within the triangle. This will be proved later.

The most elegant solution known for this problem is the following (Paul Erdös would have described it as being "from the Book"). Let P be a point within the triangle. Rotating $\triangle BPC$ about C counter-clockwise, through an angle of 60°, we get Figure 4.6.

Let the image of $\triangle BPC$ under the rotation be $\triangle XQC$, then $PB = QX$. Also, triangle PQC is equilateral (and so is triangle BXC), $\therefore$ $PC = PQ$. Therefore,

$$PA + PB + PC = AP + PQ + QX.$$

Clearly, we have $AP + PQ + QX \geq AX$, with equality iff the points A, P, Q, X are collinear (in that order). So,

$$PA + PB + PC \geq AX,$$

with equality iff A, P, Q, X are collinear.

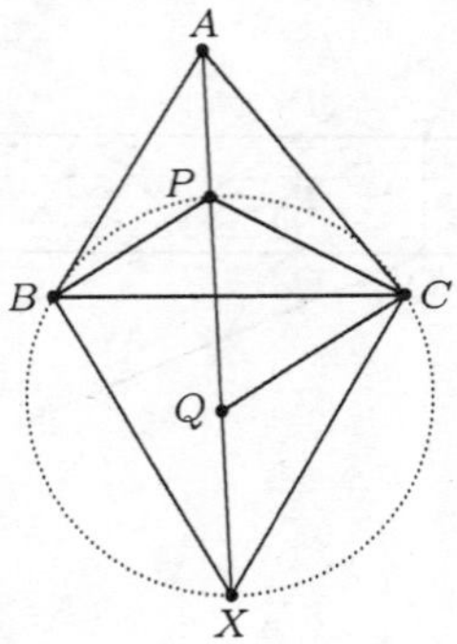

Figure 4.7. *Minimizing $AP + BP + CP$*

Now observe that if A, P, Q and X are indeed collinear, then $\angle XQC = 180° - 60° = 120°$, $\therefore \angle BPC = 120°$. Also, $\angle CPA = 180° - 60° = 120°$, $\therefore \angle APB = 120°$ too. This means that $PA + PB + PC = AX$ iff angles BPC, CPA and APB are all equal to 120°.

So, we must locate an interior point P such that angles BPC, CPA and APB are all 120°. This *is* possible, and may be done as follows (see Figure 4.7). We construct an equilateral triangle BXC with BC as base (with X, A on opposite sides of BC). Next, we draw the circumcircle Γ of triangle BXC and join AX. The point where AX intersects the portion of Γ within the triangle is the required point P. This is clear, since $\angle BPC = 120°$ and $\angle XPC = 60°$, implying that $\angle CPA = \angle APB = 120°$.

The solution technique used above yields the following important corollary. *Let $\triangle ABC$ be equilateral, and let P be any point in the plane of the triangle. Then,*

$$PB + PC \geq PA,$$

and equality holds iff P lies on the minor arc $\overset{\frown}{BC}$ of the circumcircle Γ of $\triangle ABC$. [See Figures 4.8, 4.9.]

For the proof, we apply a rotation through 60° about C as centre, in a *clockwise* direction. This takes B to A (since $\triangle ABC$ is equilateral) and P to Q, say. Then $\triangle QPC$ is equilateral, so $PC = PQ$, and since $PB = QA$ we get $PB + PC = PQ + QA \geq PA$, by the triangle inequality; $\therefore PB + PC \geq PA$.

Equality holds iff Q lies on line segment PA. This would make $\angle APC = 60°$, implying that P lies on major arc $\overset{\frown}{AC}$ of Γ (the arc bounded by A and C and containing vertex B). The same argument

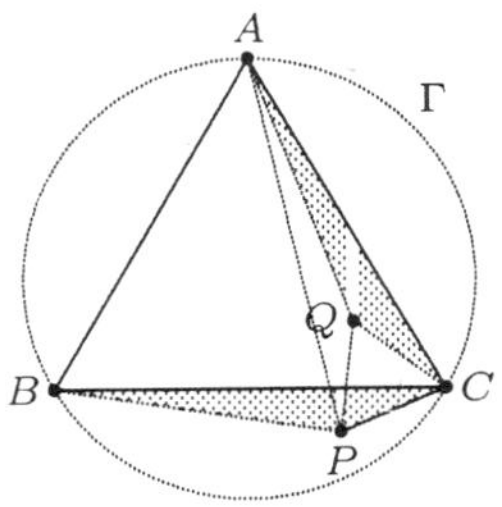

Figure 4.8.

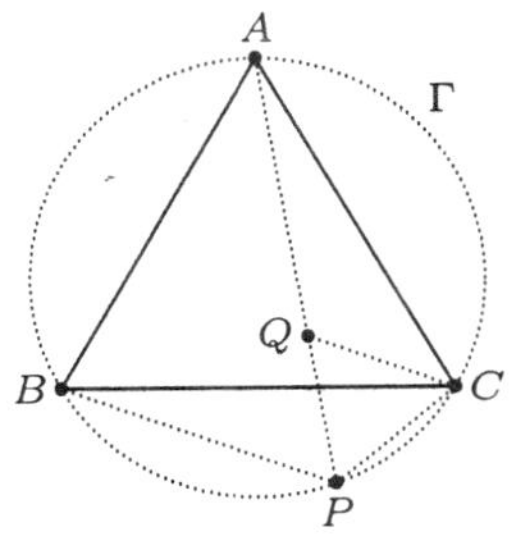

Figure 4.9.

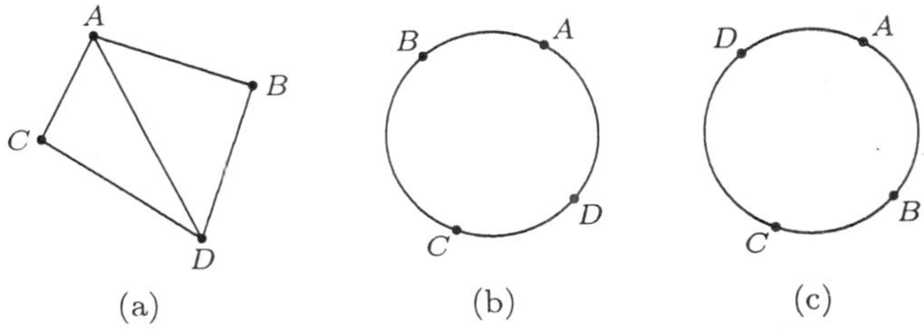

Figure 4.10. *Generalized version of Ptolemy's theorem*

repeated with the roles of B and C being interchanged, and "clockwise" replaced by "counter-clockwise" (that is, we perform a rotation about B as the centre, through $60°$, counter-clockwise), leads to the conclusion that P lies on the major arc $\overset{\frown}{AB}$. Combining the two deductions, we see that $PB + PC = PA$ iff P lies on minor arc $\overset{\frown}{BC}$ of Γ. □

Comments on Ptolemy's theorem The corollary stated above is usually proved via recourse to a much more powerful result—the generalized version of Ptolemy's theorem: *If A, B, C, D are coplanar points, then*

$$AB \cdot CD + AD \cdot BC \geq AC \cdot BD,$$

with equality iff the points A, B, C, D are concyclic and occur in that order on a circle (clockwise or counter-clockwise). [See Figures 4.10(a), (b), (c).]

Example. Let the points A, B, C and D have coordinates as follows: $A\,(0,0)$, $B\,(6,0)$, $C\,(8,8)$, $D\,(0,5)$; then the distances AB,

$BC, \ldots, CD$ are given by: $AB = 6$, $AC = \sqrt{128}$, $AD = 5$, $BC = \sqrt{68}$, $BD = \sqrt{61}$, $CD = \sqrt{73}$; so $AB \cdot CD + AD \cdot BC = 92.49$ and $AC \cdot BD = 88.36$. Observe that the first quantity exceeds the second.

Here is how Ptolemy's theorem implies the corollary. We apply the theorem to the four points P, B, A, C and get

$$PB \cdot AC + PC \cdot AB \geq PA \cdot BC,$$

or $PB + PC \geq PA$, since AB, AC, BC are equal. Equality holds iff P, B, A, C are concyclic, with the points occurring in that order on a circle; that is, iff P lies on the minor arc $\overset{\frown}{BC}$ of the circumcircle Γ of $\triangle ABC$.

The corollary may be applied to Problem 3 as follows. Let BXC be an equilateral triangle constructed on side BC of the given triangle as the base, with X and A on opposite sides of BC (see Figure 4.11). Let Γ be the circumcircle of $\triangle BXC$, and let P be any point within the triangle. Then, we have

$$PB + PC \geq PX,$$

with equality iff P lies on minor arc $\overset{\frown}{BC}$ of Γ. Next,

$$AP + PX \geq AX,$$

with equality iff A, P, X are collinear. Therefore, combining the two statements,

$$PA + PB + PC \geq AX,$$

with equality iff P lies on arc $\overset{\frown}{BC}$ and A, P, X are collinear; i.e., iff P is the point where AX intersects the minor arc $\overset{\frown}{BC}$ of Γ. So, we reach the same conclusion as we did earlier.

As a related fact, we note here that if the equilateral triangles BCX, CAY and ABZ are constructed on the sides BC, CA and AB of any triangle ABC, then the three line segments AX, BY and CZ are of equal length, and they concur at a point P_0, the *Fermat–Torricelli point* of the triangle. If no angle of the triangle exceeds 120°, then P_0 lies within the triangle, and this is where the quantity $PA + PB + PC$ achieves its least value. (If some angle equals 120°, then P_0 coincides with the vertex corresponding to that angle.)

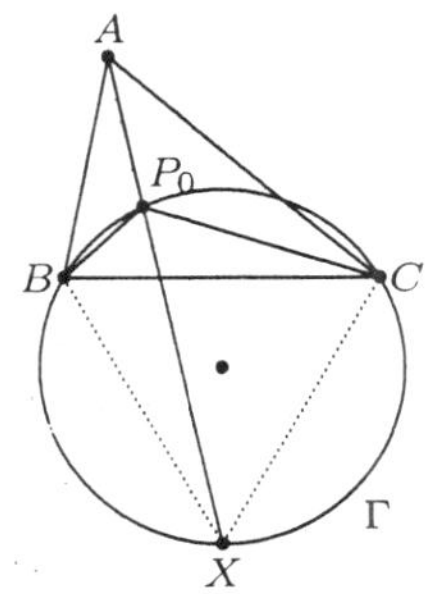

Figure 4.11.

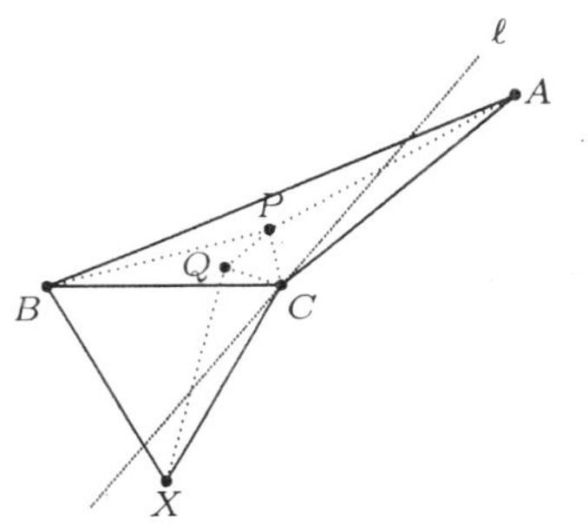

Figure 4.12.

We do not discuss the proof of Ptolemy's theorem here, as that would take us too far afield. One of the nicest proofs uses *inversion*, a sophisticated geometrical transformation of great power; another proof uses complex numbers.

⋆ ⋆ ⋆

We now explain why we needed to specify that no angle of the triangle must exceed 120°. Suppose that in $\triangle ABC$, angle C exceeds 120°. We shall show that in this case the optimizing point is vertex C. We use the corollary to Problem 1, above.

We construct as earlier an equilateral triangle BXC, with X and A on opposite sides of BC. Let P be any point within the triangle. Since angle BCA exceeds 120°, at least one of the angles BCP and PCA must exceed 60°. We assume, without loss of generality, that $\angle BCP > 60°$ (see Figure 4.12).

Let the image of P under a counter-clockwise rotation about C as the centre, through 60°, be Q. Then Q falls within the arms of $\angle BCA$. Also, $PB = QX$ and $PC = PQ$. Observe that

$\angle XCB + \angle BCA > 180°$. Let ℓ denote the external angle bisector of the angle XCA.

It cannot happen that P and Q *both* lie on the same side of ℓ as X and A. If this were to happen, then either P and Q would both lie within the angle enclosed by ℓ and CA, in which case we would have $\angle QCP < 30°$; or P would lie in the angle enclosed by ℓ and CA, and Q would lie in the angle enclosed by ℓ and CX; but these cannot be, as $\angle QPC = 60°$. Suppose that Q lies on the "far side" of ℓ relative to X. We now have, via the corollary to Problem 1,

$$XQ + QA > XC + CA, \quad \therefore \quad XQ + QP + PA > XC + CA,$$

and therefore, $PA + PB + PC > CB + CA$, since $XC = CB$. So, the optimizing point in this case is C itself. □

4.6 Problems

4.6.1 Let $\triangle ABC$ be acute-angled, and let X be a variable point on side BC. From X, perpendiculars XY and XZ are dropped to sides AB and AC, respectively. Where should X be located on BC if YZ is to have the least length possible?

HINT. Use the fact that quadrilateral $AYXZ$ is cyclic.

4.6.2 Let $\triangle ABC$ be acute-angled, and let X be a variable point on side BC. Let AX be joined and let the incircles of $\triangle ABX$ and $\triangle ACX$ be drawn. Let the two circles touch BC at Y and Z, respectively. Where should X be located on BC if YZ is to have the least length possible?

4.6.3 Let r, R denote, respectively, the radii of the incircle and the circumcircle of a triangle. Prove that $R \geq 2r$.

[*] Formulate and prove a 3-D analogue of this assertion which is valid for an arbitrary tetrahedron.

4.6.4 Let m_a, m_b and m_c be the lengths of the medians of $\triangle ABC$, and let p be its perimeter. Prove that

$$\frac{3p}{4} < m_a + m_b + m_c < p.$$

4.6.5 We offer here a pretty application of the result proved in Section 4.5 (Problem 1, Figure 4.3).

An *ellipse* can be defined[2] in the following manner. Let A and B be given points, and let k be a given distance, $k > AB$. Let $\mathcal{E}$ be the locus defined thus:

$$\mathcal{E} = \{P : AP + BP = k\}.$$

Then $\mathcal{E}$ is an ellipse with foci at A and B. Its major axis has a length of $2k$, and its eccentricity is given by $e = AB/k$.

Use this definition, together with the result of Problem 1 of Section 4.5, to show that a ray of light emanating from A and proceeding in any direction whatsoever will be reflected by the ellipse towards B. (The ellipse is taken to be a smooth, reflecting surface.)

4.6.6 Let h be the largest altitude of $\triangle ABC$, and let R and r denote, respectively, the circumradius and inradius of the triangle. Prove that $R + r \leq h$. (This result is due to Paul Erdös.)

4.6.7 Triangle ABC is right-angled at A; D is the foot of the altitude through A, F is the foot of the internal bisector of angle ABC, and E is the point where BF and AD meet. Prove that $DC > 2\,EF$.

4.6.8 (From IMO 1988, Canberra, Australia) The triangle ABC is right-angled at A, and AD is an altitude; I and J denote, respectively, the incentres of $\triangle ABD$ and $\triangle ACD$, and the line through I and J meets AB and AC at K and L respectively. Prove that the area of $\triangle AKL$ is, at the most, $1/2$ the area of $\triangle ABC$.

4.6.9 Let P be an interior point of the tetrahedron $ABCD$. Let F_a, F_b, F_c and F_d refer to the areas of the faces opposite vertices A, B, C and D, respectively. Prove that:

$$PA \cdot F_a + PB \cdot F_b + PC \cdot F_c + PD \cdot F_d \geq 9 \cdot \text{Vol}(ABCD),$$

where 'Vol' means 'Volume'. (Is there a 2-D version of this result?)

---

[2] There are other definitions, e.g., in terms of the focus and directrix.

4.6.10 Given a convex quadrilateral $ABCD$, let R_A, R_B, R_C, R_D denote, respectively, the circumradii of $\triangle BCD$, $\triangle CDA$, $\triangle DAB$, $\triangle ABC$. Show that if $\angle A + \angle C > \angle B + \angle D$, then $R_A + R_C > R_B + R_D$, and conversely.

4.7 Concluding remarks

As the rich variety of problems and examples presented above will suggest, inequalities form a vast and rich topic; new results are found on a regular basis. Reference [4], published in 1989, illustrates the richness and depth of the topic—it is a massive tome, filled with discoveries made in the 1970s and 1980s.

It seems rather a pity that inequalities are seldom encountered at the school level; one gets to know their rich potential only much later. This is unfortunate, because there are so many nice results accessible to a younger audience. We hope that the exposure in this chapter to some well-known and beautiful inequalities will stimulate the interest of the young readers in the subject.

We conclude the chapter with some challenging problems.

Exercises

4.7.1 Prove that for all positive integers n, we have the inequality

$$\left(1+\frac{1}{n}\right)^n < 3.$$

Use this result to conclude that for positive integers n, the largest value taken by $\sqrt[n]{n}$ is $\sqrt[3]{3}$.

4.7.2 For which positive integer n does the quantity

$$\sqrt[n]{n+1000000}$$

take its largest value?

4.7.3 For positive integers n, let a_n and b_n be defined as follows:

$$a_n = \left(1+\frac{1}{n}\right)^n, \quad b_n = \left(1+\frac{1}{n}\right)^{n+1}$$

Show that $a_n < a_{n+1}$ and that $b_n > b_{n+1}$ for all n.

COMMENT. This means that $\{a_n\}$ forms an *increasing sequence* while $\{b_n\}$ forms a *decreasing sequence*. Also, $a_n < b_n$ for all n (trivially so), and when n is infinitely large, $a_n \approx b_n$ (because $1 + 1/n \approx 1$). We express this by writing, in shorthand: $a_\infty = b_\infty = L$, say. We now have the following:

$$a_1 < a_2 < \cdots < a_\infty = L = b_\infty < \cdots < b_2 < b_1;$$

and

$$\left(1 + \frac{1}{m}\right)^m < L < \left(1 + \frac{1}{n}\right)^{n+1} \quad \text{for all } m, n \in \mathbf{N}.$$

4.7.4 Show that the following inequality holds for all integers $n \geq 6$:

$$\left(\frac{n}{3}\right)^n < n! < \left(\frac{n}{2}\right)^n$$

(Here $n! = 1 \times 2 \times 3 \times \cdots \times (n-1) \times n$.)

COMMENT. It can be shown that for large n the following approximation is an excellent one:

$$n! \approx \sqrt{2\pi n}\left(\frac{n}{e}\right)^n,$$

where $e \approx 2.71828$ is *Euler's constant*. This is called *Stirling's approximation*, and it immediately yields the inequalities listed above. Your task, however, is to find elementary proofs for these inequalities. For comparison, we give below the values of the various functions when $n = 100$.

- $100! \approx 9.333 \times 10^{157}$;
- $(100/3)^{100} \approx 1.94 \times 10^{152}$;
- $(100/2)^{100} \approx 7.889 \times 10^{169}$;
- $(100/e)^{100} \approx 3.72 \times 10^{156}$;
- $\sqrt{200\pi} \times (100/e)^{100} \approx 9.325 \times 10^{157}$.

4.7.5 $\boxed{*}$ Show that for $n \geq 1$, we have the inequality

$$\sqrt[n]{n} < 1 + \sqrt{\frac{2}{n}}.$$

In fact, the following inequality also holds but is harder to prove: $\sqrt[n]{n} < 1 + 1/\sqrt{n}$.

4.7.6 [*] (Continuation) Show that for $n \geq 1$, we have the inequality

$$\sqrt[n]{n} < 1 + \frac{2}{\sqrt{n}} - \frac{2}{n}.$$

The following inequality holds but is harder to prove:

$$\sqrt[n]{n} < 1 + \frac{1}{\sqrt{n}} - \frac{1}{3n}.$$

4.7.7 Let $a_1, a_2, \ldots, a_n$ be positive real numbers with sum s. Show that

$$a_1a_2 + a_2a_3 + \cdots + a_{n-1}a_n \leq \frac{s^2}{4}.$$

When does equality hold? (The sum on the left is $\sum_{i=1}^{n-1} a_ia_{i+1}$; it has $n-1$ terms in all.)

4.7.8 For positive integers n, let s_n be given by

$$s_n = 1 + \frac{1}{2^2} + \frac{1}{3^2} + \cdots + \frac{1}{n^2}.$$

Show that we have the following successively better bounds on s_n (valid for all n):

(a) $s_n < 2$;

(b) $s_n < \dfrac{61}{36} \approx 1.694$;

(c) $s_n < \dfrac{241}{144} \approx 1.674$.

Find a bound superior to the one in (c).

4.7.9 For positive integers n, let t_n be given by

$$t_n = 1 + \frac{1}{2^3} + \frac{1}{3^3} + \cdots + \frac{1}{n^3}.$$

Show that $t_n < 5/4$ for all n.

4.7.10 Show that $a_n < 1/3$ for all positive integers n, where

$$a_n = \left(1 - \frac{1}{2}\right)^2 \left(1 - \frac{1}{4}\right)^2 \cdots \left(1 - \frac{1}{2n}\right)^2 \cdot n.$$

4.7.11 $\boxed{*}$ Let a, b, c be positive real numbers. Show that

$$\frac{a+\sqrt{ab}+\sqrt[3]{abc}}{3} \le \sqrt[3]{a\frac{(a+b)}{2}\frac{(a+b+c)}{3}}.$$

Establish when equality holds.

4.7.12 $\boxed{*}$ Let ABC and $A'B'C'$ be two triangles with sides a, b, c and a', b', c'. Let the area of $\triangle ABC$ be F and let the circumradius of $\triangle A'B'C'$ be R'. Show that

$$\frac{a^2}{a'}+\frac{b^2}{b'}+\frac{c^2}{c'} \ge \frac{4F}{R'}.$$

4.7.13 Let P be a point interior to $\triangle ABC$, and let B_1 and C_1 represent arbitrary points on AB and AC, respectively. Let y and z represent the distances of P from AC and AB respectively, and let u represent the distance PA. Prove that

$$B_1C_1 \cdot u \ge AB_1 \cdot z + AC_1 \cdot y.$$

4.7.14 $\boxed{*}$ **Erdös–Mordell Inequality** Let P be a point interior to $\triangle ABC$. Let u, v and w represent the distances PA, PB and PC, and let x, y and z represent the perpendicular distances of P from the sides BC, CA and AB (respectively, in each case). Show that $u+v+w \ge 2(x+y+z)$. When does equality hold?

HINT. Use the preceding result with suitable choices for B_1 and C_1 (let B_1, C_1 be the images of C, B under reflection in the bisector of $\angle BAC$). The resulting expressions, after some manipulation, yield the Erdös–Mordell inequality. Other choices yield more such inequalities.

EXAMPLE. By choosing B_1, C_1 to be B, C, we obtain $au \ge by+cz$. Three such inequalities, when added together, yield:

$$au+bv+cw \ge 2(ax+by+cz).$$

4.7.15 $\boxed{*}$ Let P be a point interior to an acute-angled $\triangle ABC$ of unit circumradius. Show that

$$AP \cdot BP \cdot CP < \frac{32}{27}.$$

(This result too is due to Paul Erdös.)

4.8 References

[1] Beckenbach, E and Bellman, R. *An Introduction to Inequalities*, MAA.

[2] Bottema, O *et al.* *Geometric Inequalities*, Wolters-Noordhoff.

[3] Korovkin, P P. *Inequalities*, MIR Publishers, 1983.

[4] Littlewood, J E, Hardy, G H and Polya, G. *Inequalities*, Cambridge.

[5] Mitrinović, D S, Pečarić, J E and Volenec, V. *Recent Advances in Geometric Inequalities*, Kluwer.

Chapter 5

Chessboard Frolic

The first point we must make is that this chapter is not about chess but about the chessboard! Chess problems form a genre by themselves and are immensely popular; they are featured regularly in magazines and newspapers. Problems involving the chessboard itself are less well known, but nonetheless popular. Here, for instance, are two problems from England's greatest puzzlist, Henry Ernest Dudeney (1847–1930). They are Problems 288 and 294 from his well-known book, *Amusements in Mathematics*. The reader is invited to tackle them before proceeding any further.

In Problem 288, Dudeney asks for the number of ways in which the chessboard may be divided into two parts of the same size and shape, by cuts along the lines dividing the squares. (Reflections and rotations are not counted as different.) For a 2×2 board there is clearly just one way of doing this (by a straight line running down the centre). For a 4×4 board we find, after some experimentation, that there are six possible ways (two are shown in Figure 5.1; find the other four!). Dudeney then poses the problem of finding the corresponding number for a 6×6 board.

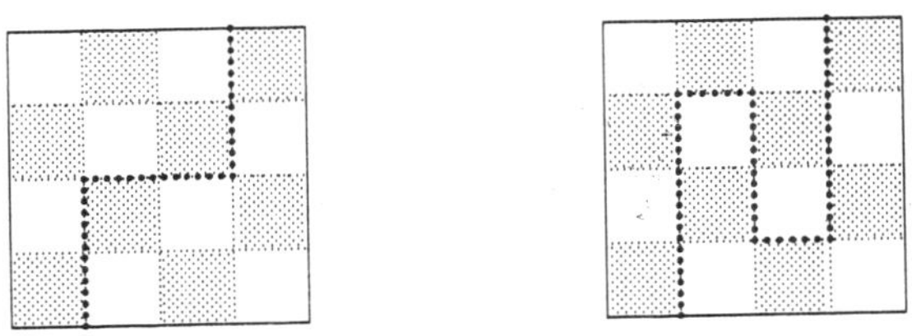

Figure 5.1. *Cutting a 4×4 board into two congruent parts*

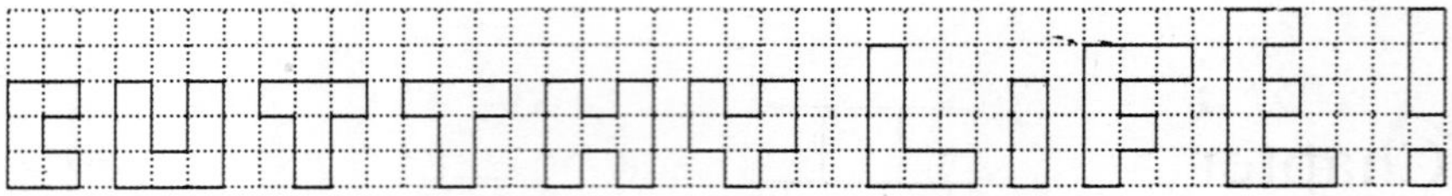

Figure 5.2. *Cut Thy Life!*

(a) (b)

Figure 5.3. *A domino and a forbidden shape.*

In Problem 294, Dudeney describes how he once set himself "The task of so dissecting an ordinary chessboard into letters of the alphabet that they would form a complete sentence", and came up with the solution shown, with the enigmatic instruction 'Cut Thy Life' (see Figure 5.2). (His sentence had four full stops, which we have reassembled into an exclamation mark!) Dudeney now asks: *Can you fit the pieces together and form the original* 8×8 *board?*

⋆ ⋆ ⋆

We shall see in this chapter that a chessboard is an extremely convenient object for a problemist: many pretty problems can be posed using its lines and squares, and many principles of combinatorics can be illustrated very simply using a chessboard.

We shall start with a study of *polyominoes*, a class of objects first popularized by the mathematician Solomon Golomb.

5.1 Polyominoes

A shape of the type shown in Figure 5.3(a), made up of two squares that share an edge, is known as a *domino*. There is clearly only one domino shape possible. The shape shown in (b) is not acceptable—no gluing is allowed!

If three squares have to be stuck together to make a 2-dimensional figure, we can do so in two ways, as shown in Figure 5.4. Note that edges must get joined to edges, *whole*. These

Figure 5.4. *Straight tromino and L-tromino.*

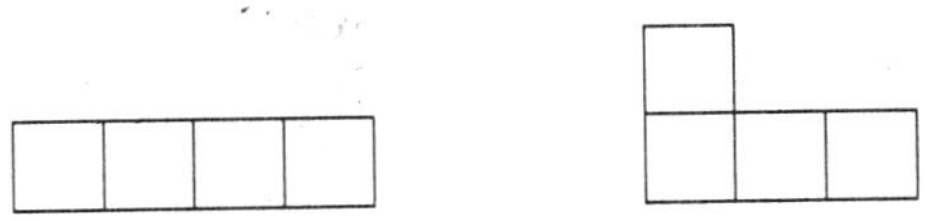

Figure 5.5. *Straight tetromino and L-tetromino.*

shapes are the *trominoes*, and only two distinct trominoes are possible.

Shapes made with four squares are the *tetrominoes*.

To continue, shapes made using five squares and six squares are known respectively as *pentominoes* and *hexominoes*. Examples are shown in Figure 5.6.

In general, an n-omino (a *polyomino* with n squares) is a shape cut from a chessboard, using the lines of the board to make the cuts, containing n squares and shaped so that a rook ("castle" to a chess player) can travel from any of its squares to any other square. In brief: *an n-omino is a rook-wise connected set of n cells of a chessboard.*

EXERCISES

5.1.1 By experimenting on squared paper, list all possible tetromino shapes (there are five shapes in all).

5.1.2 Make a list of all possible pentomino shapes.

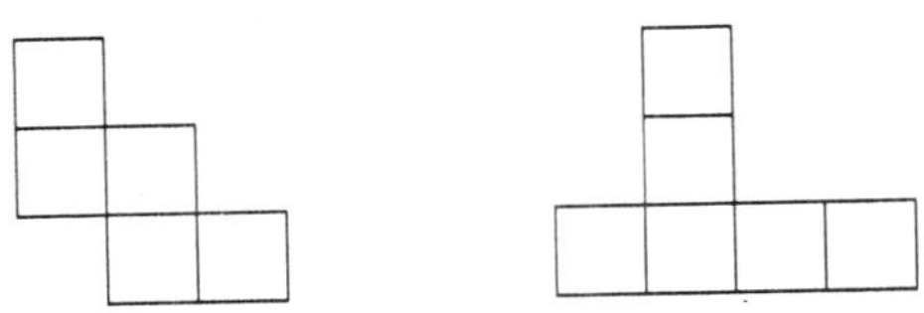

Figure 5.6. *A pentomino and a hexomino.*

Figure 5.7. *The L-tetromino.*

REMARK 1. *What does "all possible shapes" mean?* We need to explain what "all possible shapes" means. In geometric language, our task is to prepare a list of polyomino shapes of the given size such that no two of the shapes are congruent to one another, and every possible such shape is congruent to some shape on our list.

For example, of the two shapes shown in Figure 5.7, we include only one in our collection (either one will do), as the two shapes are congruent to one another. For obvious reasons, we call the shape a *L-tetromino*.

REMARK 2. *Counting the polyominoes.* Let p_n denote the number of polyominoes with n squares ($p_1 = 1$, $p_2 = 1$, $p_3 = 2$, $p_4 = 5$, ...). It is not hard to guess that p_n grows rapidly with n. The reader may, however, be surprised to learn that the computation of p_n for large n is considered a very difficult task. At present, no simple formula is known for p_n, nor any simple algorithm that can compute values of p_n for large n, say p_{10000} or p_{20000}.

5.2 Tiling

The problems listed below deal with *tilings*. A set of shapes $\mathcal{S}_i$ is said to *tile* a given larger shape $\mathcal{S}$, if the shapes can be placed on $\mathcal{S}$ so as to completely cover it. There should be no gaps, nor should any overlaps occur among the $\mathcal{S}_i$s; each point of $\mathcal{S}$ should be covered by one and only one $\mathcal{S}_i$. An obvious necessary condition for this is that the sum of the areas of the $\mathcal{S}_i$ should be equal to the area of $\mathcal{S}$. It is nearly as obvious, however, that this condition by itself will not be enough to ensure that a tiling exists. For the tiling to work, the shapes need to fit together in the right way—the corners of one should fit with the indentations in another, the angles should match, and so on.

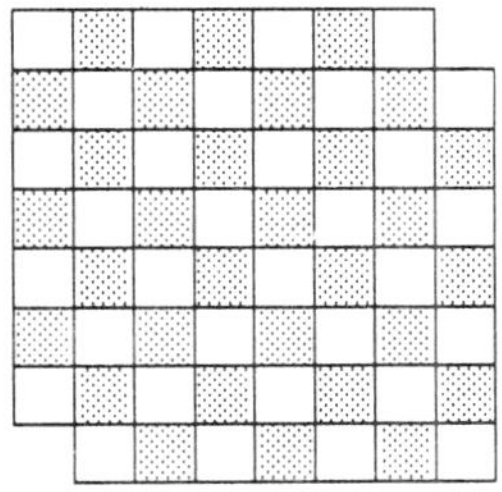

Figure 5.8. *The mutilated chessboard.*

As an example of how an area argument may be used, consider the following problem.

> **The mutilated chessboard** *A chessboard can clearly be tiled with a set of 32 dominoes. Suppose that from the chessboard we snip away two diagonally opposite corners, leaving behind 62 squares. Question: Can 31 dominoes be used to tile this shape?* [See Figure 5.8.]

The areas match ($64 - 2 = 62$, $31 \times 2 = 62$), but what if we consider the black and white squares separately?

Observe that a domino placed anywhere on the board (with its sides parallel to the sides of the board) occupies one black and one white square; this is so whether the domino is placed horizontally or vertically. So, if 31 dominoes are placed on the board without overlaps, they will cover precisely 31 black and 31 white squares. But our mutilated board has 32 squares of one colour and 30 squares of the other colour. Therefore, the tiling is not possible! □

Here the direct argument based on area does not yield anything, but by suitably colouring the shape S we make immediate progress.

The reader is urged to make sets of tetrominoes and pentominoes. They are a rich source for puzzles and problems of many kinds, as well as being lots of fun to play with!

EXERCISES

5.2.1 *The mutilated chessboard revisited.* Consider again the question of whether a chessboard from which some two squares have been snipped away can be tiled using 31 dominoes.

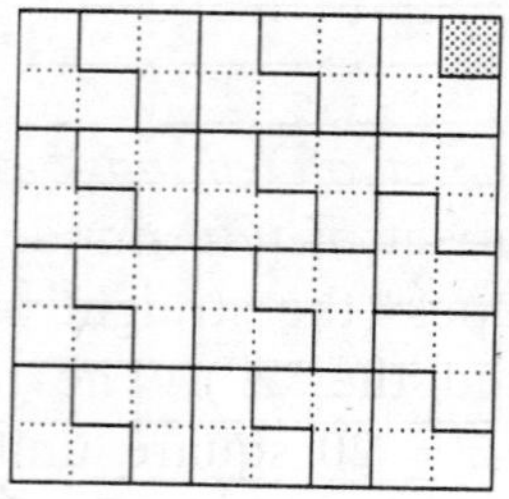

Figure 5.9. *Tiling with the L-tromino.*

Prove that if the deleted squares have *opposite colours*, then a tiling is always possible, irrespective of the precise placement of the two squares.

5.2.2 *Tiling with the L-tromino.* Next, we consider the problem of tiling a chessboard with 21 L-trominoes and 1 monomino. In terms of area we do have a match ($(21 \times 3)+1 = 64$), and indeed a tiling is possible (see Figure 5.9; the monomino is shown shaded).

The following question poses itself quite naturally: *What happens if we change the placement of the monomino to some arbitrary square? Will it always be possible to complete the tiling?* Show that this is indeed the case. That is, show that it is possible to place 21 L-trominoes on the chessboard, without overlaps, so that the uncovered square occupies any desired position.

5.2.3 *Tiling with the straight tromino.* If we attempt to tile a chessboard using 21 straight trominoes (of the 3×1 shape) plus a monomino, we soon experience difficulties. It turns out that for the tiling to be successful, there are just *four* possible squares which the monomino can occupy! Identify these squares and prove the assertion.

5.2.4 *Tiling with a 4×1 tetromino.* Let a large supply of straight (4×1) tetrominoes be given. Suppose that it is possible to tile a rectangle of dimensions $m \times n$ with these tetrominoes. Show that at least one out of m, n must be a multiple of 4.

COMMENT. We infer from this that 25 straight tetrominoes cannot tile a square of size 10×10—a far-from-obvious assertion.

5.2.5 *Tiling with the full set of tetrominoes–1.* In this problem, we use a complete set of tetrominoes, each containing one of the 5 basic shapes (the straight piece, the 2×2 square, the 'L', the 'T' and the 'Z'). The total area occupied by the pieces is $4 \times 5 = 20$ square units. We mark out on a chessboard a 4×5 rectangle. Can the rectangle be tiled using the 5-piece tetromino set?

5.2.6 *Tiling with the full set of tetrominoes–2.* Unfortunately, one cannot mark out a 2×10 rectangle on a chessboard—the board isn't big enough! However, we can work with larger chessboards. Can the 5-piece tetromino set be used to tile a 2×10 rectangle?

5.2.7 *Tilings with the full set of pentominoes.* You would have found earlier that the pentomino set comprises 12 shapes in all ($p_5 = 12$). Investigate the tilings of the following shapes using a complete set of pentominoes:

(a) a 5×12 rectangle;

(b) a 6×10 rectangle;

(c) a 4×15 rectangle;

(d) a 3×20 rectangle;

(e) a 8×8 square. (Here, four cells remain uncovered. The requirement now is that the four cells must have the shape of some tetromino.)

5.2.8 Though 3-dimensional chess has yet to be invented (the practical difficulties would seem to be considerable!), we advance now to three dimensions.

A *domino brick* is a cuboid with measurements $2 \times 1 \times 1$ (see Figure 5.10).

Using thirteen such bricks, is it possible to assemble a $3 \times 3 \times 3$ cube with a hole in its center? (Note that there is agreement on the total volume: $13 \times 2 = 26$ and $27 - 1 = 26$.)

5.2.9 Can nine L-tromino bricks be used to assemble a solid $3 \times 3 \times 3$ cube?

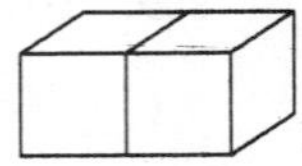

Figure 5.10. *A domino brick*

		x			c	c	c
		x			c	K	c
		x			c	c	c
		x					
x	x	R	x	x	x	x	x
		x					
		x					
		x					

Figure 5.11. *Individual coverings*

5.2.10 Make a complete list of the possible tetromino bricks. How many such bricks are there?

5.2.11 We are given a large supply of two types of bricks, of the $2 \times 2 \times 2$ and $3 \times 3 \times 3$ types, and our task is to assemble a finite number of them into a solid cuboid, of any dimension whatsoever, with no holes within it. Show that the only cuboids that can be assembled are those that can be made using just one type of brick.

5.3 Coverings and packings

We now bring in the chess pieces and consider another genre of problems defined on the chessboard.

Imagine that we had large supplies of every kind of chess piece. For each piece placed anywhere upon the board, there exists a region on the board that is 'under threat' by that piece. For example, consider Figure 5.11. For the rook (R), the under-threat region consists of the squares marked 'x', together with the square occupied by R itself, and for the king (K), the region under threat consists of the squares marked 'c', together with the square occupied by K himself.

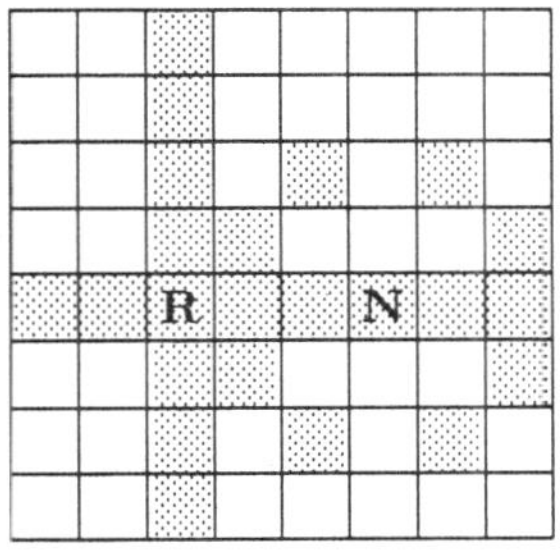

Figure 5.12. *Joint coverings.*

If two or more pieces are placed on the board, we consider the region under threat to be the collection of all squares that are under threat by at least one of the pieces (considered individually); that is, it is the *union* of the individual under-threat regions. Figure 5.12 shows an example; the under-threat region is shown shaded.

Note that the squares on the right side of the knight (N) are considered to be under threat by the rook (R). So, our definition differs from that in the actual game of chess.

A set of pieces placed on the board is said to *cover* the board if every square is under attack by some piece, and the placement itself is called a *covering*.

EXAMPLE. Let a rook be placed on each square of the main diagonal, i.e., on the squares in the ith row and i^{th} column for $i = 1, 2, \ldots, 7, 8$ (eight rooks in all). This clearly constitutes a cover.

The *covering number* of a piece is the least number k such that k units of the piece can be placed on a 8×8 board so as to completely cover it.

EXAMPLE. The covering number for a rook cannot exceed 8 (because there exists a complete cover using 8 rooks). Could the covering number be 7 or less? That is, can we cover the board using 7 (or fewer) rooks?

EXERCISES

5.3.1 Find the covering number for a rook.

5.3.2 Find the covering number for a king.

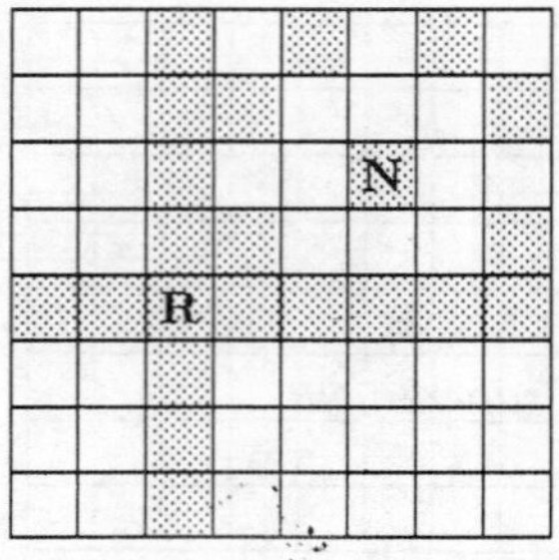

Figure 5.13. *A packing.*

5.3.3 Find the covering number for a queen.

5.3.4 Find the covering number for a knight.

5.3.5 Suppose that the size of the chessboard is 11×11. Can 5 queens cover this board?

The 'inverse' problem We now consider the "inverse" problem—that of placing a set of pieces upon the board so that *no piece is under threat from any other piece*. Such a configuration is termed a *packing*. An example is shown in Figure 5.13. (The R and N do not threaten one another.)

Analogous to the covering number of a piece, we define the *packing number* of a piece to be the largest number k such that k units of that piece can be placed on an 8×8 board, with no two of them threatening one another.

EXERCISES

5.3.6 Find the packing number for a rook.

5.3.7 Find the packing number for a queen.

5.3.8 Solve the queen-packing problem for chessboards of the following different sizes: 3×3, 4×4, 5×5, 6×6 and 7×7. What general observation can be made with regard to the $n \times n$ chessboard?

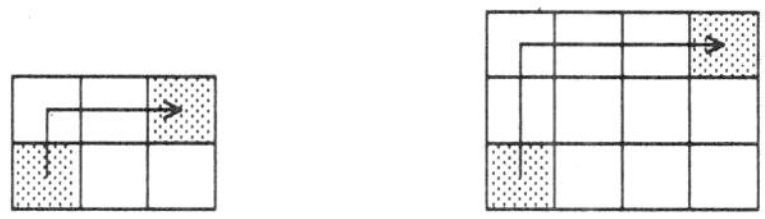

Figure 5.14. *Knight and superknight.*

5.4 Knight tours

A *knight tour* of a chessboard is a tour of the board by a knight, in which each square is visited precisely once. The tour is *closed* if, from the last square of the tour, it is possible to move to the starting square, otherwise it is *open*.

Exercises

5.4.1 Investigate the possibility of a knight tour (closed or open) on boards of sizes 3×3 and 4×4.

5.4.2 Repeat the investigation for boards of sizes 5×5 and 6×6.

5.4.3 A knight's move can be thought of as being a jump between a pair of opposite corner squares of a 2×3 rectangle, so we may refer to the knight as a "$(2, 3)$-knight". We now create a *superknight*, which jumps between a pair of opposite corner squares of a 3×4 rectangle, and refer to it as a "$(3, 4)$-knight". (See Figure 5.14.)

Investigate the possibility of a superknight tour on a 12×12 chessboard.

5.4.4 (This problem was number 3 in the International Mathematical Olympiad of 1996, held that year in Mumbai, India.)

Instead of regarding a knight's move as being between opposite corner squares of a $(2, 3)$ rectangle, we may also regard it as being a jump of length $\sqrt{5}$. Here, the 'length' of the jump is the distance between the centres of the starting square and the ending square. This allows for a generalization: we allow jumps of length $\sqrt{r}$ where r is some fixed positive integer.

Suppose that the board $ABCD$ measures 20×12, with $|AB| = 20$ and $|BC| = 12$, and suppose that our task is to

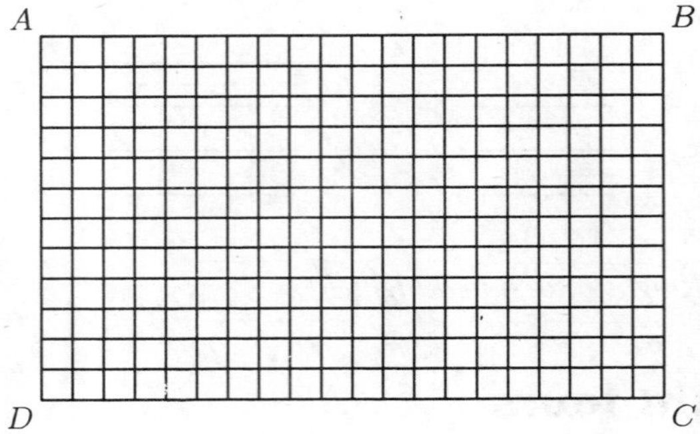

Figure 5.15. *A problem from IMO 1996*

find a sequence of jumps leading from the corner square with A as a vertex to the corner square with B as a vertex. (See Figure 5.15.)

(a) Show that if r is even or a multiple of 3, then the task cannot be accomplished.
(b) Show that if $r = 73$, then the task can be accomplished, and find a path of the required type.
(c) What happens when $r = 97$?

5.5 The Soma cube

We conclude the chapter with a brief account of a very popular puzzle, the *Soma cube*, invented by Piet Hein, a Danish writer. Consider all possible irregular shapes that can be made by gluing together four or fewer cubes of size $1 \times 1 \times 1$, how many such shapes are possible? The number is found to be seven, and the resulting shapes constitute the Soma cube (see Figure 5.16). The word 'irregular' here has the connotation that the shape must be *non-convex*. The cuboid shapes measuring $1 \times 1 \times 1$, $1 \times 1 \times 2$, $1 \times 1 \times 3$ and $1 \times 2 \times 2$, though they contain four or fewer cubes, are convex and, therefore, ruled out by this requirement.

The Soma pieces certainly pack a lot of fun in them. The author constructed a set while in high school, and it is still with him, going strong! (He got the idea from Martin Gardner's charming little book, *More Mathematical Puzzles and Diversions*.)

If we compute the total number of unit cubes used in making the seven-piece Soma set, we get a figure of 27. Rather conveniently,

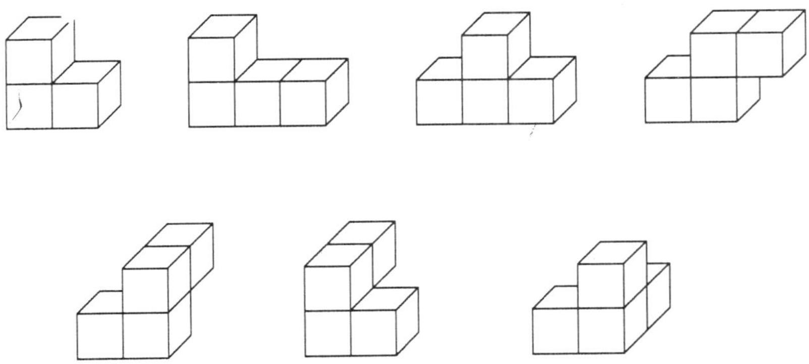

Figure 5.16. *The seven pieces of the Soma cube puzzle*

this number is equal to $3 \times 3 \times 3$. So the question of whether these pieces can be put together to form a $3 \times 3 \times 3$ cube is a meaningful one. It turns out that the construction *is* possible, and there are said to be more than 230 ways of doing so. (This is the number quoted by Martin Gardner.) After a few days of play with the pieces one acquires an excellent intuitive sense about them, and the construction of the cube can be done almost at will, starting from any arbitrary two-piece combination.

What is extremely appealing about the Soma puzzle is that a large number of pretty objects can be assembled with them. Obviously, there are also numerous objects that *cannot* be assembled, even though they happen to be made from 27 unit cubes. The proofs that these shapes cannot be assembled can be hard to find! Shown in Figure 5.17 are two shapes; the sofa can be assembled, but not the shape shown at the right. Readers may enjoy devising a proof of impossibility. Arguments based on a chessboard-type black/white colouring should work.

5.6 References

[1] Dudeney, H E. *Amusements in Mathematics*, Dover, 1970.

[2] Gardner, M. *More Mathematical Puzzles and Diversions*, Pelican, 1961.

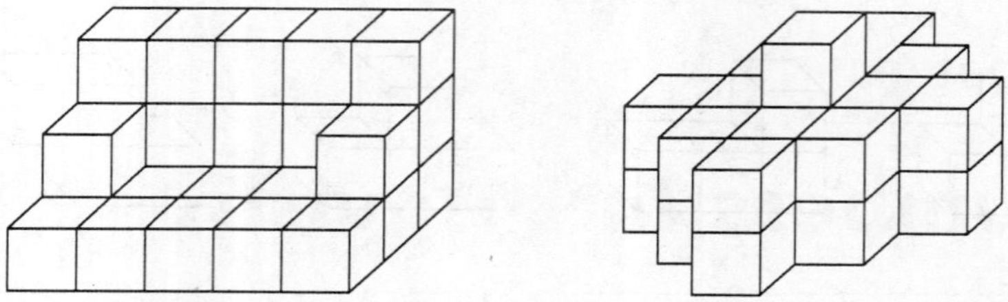

Figure 5.17. *One of these shapes cannot be made using the Soma pieces!*

[3] Golomb, S. *Polyominoes*, Sribner's, 1965.

[4] Lindgren, H. *Recreational Problems, Geometric Dissections and How to Solve Them*, Dover, 1972.

Chapter 6

Numbers In Disguise

6.1 Cryptarithms

A *cryptarithm* is an arithmetical puzzle in which each digit of an arithmetical operation (addition, subtraction, multiplication, ...) is assigned a definite letter as in a code. The problem is to break the code. We have already made a detailed study in Chapter 1 of one such problem:

$$\overline{\mathsf{ABCD}} \times 9 = \overline{\mathsf{DCBA}}$$

We decoded this thus: $\mathsf{A} = 1$, $\mathsf{B} = 0$, $\mathsf{C} = 8$, $\mathsf{D} = 9$; answer: $1089 \times 9 = 9801$.

For those who enjoy posing problems, the task of creating a good cryptarithm is a most challenging one. Several competing considerations that need to be kept in mind are given below.

- The code must be 1–1: all occurrences of a given digit must be replaced by the same letter, and different digits cannot be represented by the same letter.
- Common arithmetical conventions must be observed; e.g., the first digit of a number cannot be 0. Thus, in the problem referred to above ($\overline{\mathsf{ABCD}} \times 9 = \overline{\mathsf{DCBA}}$), we must have $\mathsf{A} \neq 0$ and $\mathsf{D} \neq 0$.
- (This is the difficult part!) The problem and code must be such that *precisely one answer* exists—i.e., there should be just one way of decoding the problem. (Mathematicians refer to such a feature as *uniqueness*.) Problems with multiple solutions are

generally not considered 'nice'. Note that this is a requirement of aesthetics, not of mathematics! It is difficult to achieve and we shall sometimes be forced to relax this requirement.

- Finally, it is also an attractive feature to select the code so that actual names or dictionary words appear in the problem, which thereby acquires a double meaning. This, too, is difficult to achieve in practice.

6.2 Notation

• **Representation of base-10 numbers** We shall use a different font when referring to the digital representation of numbers. Thus, we refer to the 2-digit number with units digit A and tens digit B as $\overline{\mathsf{BA}}$ rather than BA (which might be interpreted as B times A). Writing the number as $\overline{\mathsf{BA}}$, with a line above BA, serves to draw attention to this fact, as does the choice of a different font.

• **Symbol for congruence** We shall frequently use the symbol '$\equiv$', which stands for "is congruent to". It is defined thus: given three numbers a, b and c, with $c \neq 0$, we write

$$a \equiv b \pmod{c},$$

(to be read aloud as "a is congruent to b modulo c"; the "modulus" is c) to mean that a and b leave the same remainder when divided by c; equivalently, that $(a-b)/c$ is an integer. For example, we may write $21 \equiv 3 \pmod{6}$, or $45 \equiv 12 \pmod{11}$, or $23\frac{1}{2} \equiv 2\frac{1}{2} \pmod{11}$.

EXAMPLE. In the problem mentioned above,

$$\overline{\mathsf{ABCD}} \times 9 = \overline{\mathsf{DCBA}},$$

we have $\mathsf{D} \times 9 \equiv \mathsf{A} \pmod{10}$, because the units digit of $\mathsf{D} \times 9$ is A.

In our context, the modulus is 10 throughout (because we work in base-10), so we drop the '(mod 10)' and simply write '$\equiv$'. So, all through this chapter the notation $a \equiv b$ will mean:'$a - b$ is a multiple of 10'.

• **Symbol for the 'carries'** We need a symbol to refer to the amounts 'carried' between columns, so we number the columns thus: the units column is Column 0, the tens column is Column

1 (corresponding to $10^1 = 10$), the hundreds column is Column 2 (corresponding to $10^2 = 100$), and so on. The carryover from Column i into Column $(i+1)$ will be denoted by c_{i+1}. Thus, in the multiplication shown below,

$$\begin{array}{r} 1\ 2 \\ \times\ 7 \\ \hline 8\ 4 \\ \hline \end{array}$$

we have $c_1 = 1$ (" $7 \times 2 = 14$, carry 1").

The purpose of this notation is to make the explanations transparent. The reader will find that it helps not only in *explaining* the solution of a cryptarithm, but in *solving* it too. This is typical of a good mathematical notation—it invariably helps in analysis and problem solving.[1]

6.3 Two solved examples

At this point, the reader may want to study the solution to the '$\overline{\mathsf{ABCD}} \times 9 = \overline{\mathsf{DCBA}}$' problem, referred to above, to see how a typical cryptarithm is unraveled. Here are two additional (and rather more difficult) examples.

6.3.1 Send more money!

This is surely one of the most celebrated of all cryptarithms! It is a most attractive and intricate problem, as well as instructive to solve. We give the full solution below.

The problem is to decode the addition sum,

$$\overline{\mathsf{SEND}} + \overline{\mathsf{MORE}} = \overline{\mathsf{MONEY}},$$

which is written in columnar form as follows:

$$\begin{array}{ccccc} & \mathsf{S} & \mathsf{E} & \mathsf{N} & \mathsf{D} \\ & \mathsf{M} & \mathsf{O} & \mathsf{R} & \mathsf{E} \\ \hline \mathsf{M} & \mathsf{O} & \mathsf{N} & \mathsf{E} & \mathsf{Y} \\ \hline \end{array}$$

Note that the second digit in $\overline{\mathsf{MONEY}}$ is the letter 'O' and not a zero! In all, 8 digits have been used in the problem, corresponding to the letters $\mathsf{S, E, N, D, M, O, R, Y}$.

[1] Devising a good mathematical notation is to be regarded as part of mathematics.

This is a complex problem and, to cope with its complexity, good notation is needed. In the analysis below, we make extensive use of the symbols for congruence and "carries" (defined above).

The largest possible sum of two 4-digit numbers is

$$9999 + 9999 = 19998,$$

so $\overline{\mathsf{MONEY}} \leq 19998$ and therefore, $\mathsf{M} < 2$. It follows that $\mathsf{M} = 1$ (since $\mathsf{M} \neq 0$) and $c_4 = 1$.

The triple occurrence of E and the double occurrences of N, O, M are important clues, which we must use effectively. Note that the carryovers c_i are all 0 or 1 (because the addition is a 2-line one).

The following relations are immediate:

$$\begin{aligned} \mathsf{D} + \mathsf{E} &\equiv \mathsf{Y}, \\ c_1 + \mathsf{N} + \mathsf{R} &\equiv \mathsf{E}, \\ c_2 + \mathsf{E} + \mathsf{O} &\equiv \mathsf{N}, \\ c_3 + \mathsf{S} + 1 &\equiv \mathsf{O}. \end{aligned}$$

(Indeed, $c_3 + \mathsf{S} + 1 = 10 + \mathsf{O}$, since $c_4 = 1$.) These relations may also be written as follows:

$$\mathsf{D} + \mathsf{E} = 10c_1 + \mathsf{Y}, \tag{3.1}$$

$$c_1 + \mathsf{N} + \mathsf{R} = 10c_2 + \mathsf{E}, \tag{3.2}$$

$$c_2 + \mathsf{E} + \mathsf{O} = 10c_3 + \mathsf{N}, \tag{3.3}$$

$$c_3 + \mathsf{S} + 1 = 10 + \mathsf{O}. \tag{3.4}$$

From (3.4), we see that $\mathsf{S} \geq 8$; for if $\mathsf{S} \leq 7$, then we would have $c_3 + \mathsf{S} + 1 \leq 1 + 7 + 1 = 9$, but $10 + \mathsf{O} \geq 10$. Therefore, $\mathsf{S} = 8$ or 9.

Equation (3.4) also yields: $c_3 + \mathsf{S} + 1 \leq 1 + 9 + 1 = 11$, therefore, $10 + \mathsf{O} \leq 11$ or $\mathsf{O} \leq 1$. This means that $\mathsf{O} = 0$ or 1. Since the digit '1' has already been assigned to M, it follows that $\mathsf{O} \neq 1$; therefore, $\mathsf{O} = 0$.

We now update our records (the entries in the top row indicate the column numbers, and those in the bottom row indicate the 'carries'), incorporating what has been uncovered till now.

4	3	2	1	0
	S	E	N	D
	1	0	R	E
1	0	N	E	Y
1	c_3	c_2	c_1	–

Scanning column 2 next, we see that there must be a carry into column 2 from column 1 (since $\mathsf{N} \neq \mathsf{E}$). Therefore,

$$c_2 = 1, \quad \mathsf{N} = \mathsf{E} + 1.$$

The relations (3.1)–(3.4) are now updated:

$$\begin{aligned} \mathsf{D} + \mathsf{E} &= 10c_1 + \mathsf{Y}, &(3.5)\\ c_1 + (\mathsf{E}+1) + \mathsf{R} &= 10 + \mathsf{E}, &(3.6)\\ 1 + \mathsf{E} + \mathsf{O} &= 10c_3 + (\mathsf{E}+1), &(3.7)\\ c_3 + \mathsf{S} &= 9. &(3.8) \end{aligned}$$

From (3.7), we see that $c_3 = 0$, and therefore from (3.8) we have $\mathsf{S} = 9$. A further update is now needed.

4	3	2	1	0
	9	E	E+1	D
	1	0	R	E
1	0	E+1	E	Y
1	0	1	c_1	–

Also,

$$\begin{aligned} \mathsf{D} + \mathsf{E} &= 10c_1 + \mathsf{Y}, &(3.9)\\ c_1 + \mathsf{R} &= 9. &(3.10) \end{aligned}$$

If c_1 were 0, (3.10) would yield $\mathsf{R} = 9$, which is not possible (since $\mathsf{S} = 9$); therefore, $c_1 = 1$ and $\mathsf{R} = 8$. Equation (3.9) now reads:

$$\mathsf{D} + \mathsf{E} = 10 + \mathsf{Y}. \qquad (3.11)$$

Note that $\mathsf{D}, \mathsf{E}, \mathsf{Y}$ cannot take any of the values 0, 1, 8 or 9 (they have been "used up"). In what way can $\mathsf{D} + \mathsf{E}$ exceed 11 while avoiding these digits? Experimentation yields the following possibilities:

- $10 + \mathsf{Y} = 12$, with $(\mathsf{D}, \mathsf{E}, \mathsf{Y}) = (5, 7, 2)$ or $(7, 5, 2)$.
- $10 + \mathsf{Y} = 13$, with $(\mathsf{D}, \mathsf{E}, \mathsf{Y}) = (6, 7, 3)$ or $(7, 6, 3)$.

(Note that $10 + \mathsf{Y}$ cannot be 14 or more, as $14 = 7 + 7$, $6 + 8$ or $5 + 9$, each of which is illegal as per our rules; and likewise for values greater than 14.)

Some of the possibilities listed are quickly eliminated: thus, $\mathsf{E} = 7$ would lead to $\mathsf{N} = \mathsf{E} + 1 = 8$, but we know that $\mathsf{R} = 8$. This shuts out the two possibilities $(5, 7, 2)$ and $(6, 7, 3)$ for $(\mathsf{D}, \mathsf{E}, \mathsf{Y})$, leaving only $(7, 5, 2)$ and $(7, 6, 3)$ to consider.

The second possibility gives $\mathsf{E} = 6$, $\mathsf{N} = \mathsf{E} + 1 = 7$ and also $\mathsf{D} = 7$, so this too is illegal. That leaves just one possibility: $(\mathsf{D}, \mathsf{E}, \mathsf{Y}) = (7, 5, 2)$, and now we find that we have indeed zeroed in on the solution:

$$\begin{array}{ccccc} & 9 & 5 & 6 & 7 \\ & 1 & 0 & 8 & 5 \\ \hline 1 & 0 & 6 & 5 & 2 \\ \hline \end{array}$$

The code is thus:

$$\mathsf{S} = 9,\ \mathsf{E} = 5,\ \mathsf{N} = 6,\ \mathsf{D} = 7,\ \mathsf{M} = 1,\ \mathsf{O} = 0,\ \mathsf{R} = 8,\ \mathsf{Y} = 2,$$

and this is the *only* answer.

A pleasing achievement indeed! One must marvel just as much at the skill that went into the creation of this problem as at the delicate play of logic that allowed for its unraveling.

6.3.2 Two times two equals three

The second problem we solve involves multiplication, and it is decidedly simpler than the first one. The problem is to decode the following:

$$\overline{\mathsf{TWO}}^2 = \overline{\mathsf{THREE}}.$$

We reason as follows.

- Since $\overline{\mathsf{TWO}}^2$ has 5 digits, it follows that $\overline{\mathsf{TWO}} \le 316$ (because $\sqrt{10^5}$ lies between 316 and 317). Therefore, T is either 1, 2 or 3.
- Since the left-most digit of $\overline{\mathsf{TWO}}^2$ is the same as that of $\overline{\mathsf{TWO}}$, namely T, we must have $\mathsf{T} = 1$. (If T were 2, then $\overline{\mathsf{TWO}}^2$ would exceed 40000 and we would have $\mathsf{T} \ge 4$; and if T were 3, then $\overline{\mathsf{TWO}}^2$ would exceed 90000 and we would have $\mathsf{T} = 9$.)
- The units digit of a square when written in base-10 must be one of the following: 0, 1, 4, 5, 6, 9. Since O and E are distinct digits, neither can be 0. Therefore, E is 1, 4, 5, 6 or 9.
- We now use the following property of squares written in base-10: *The digit in the tens place of an odd square is never odd.* (For example, consider the odd squares $123^2 = 15129$ and $137^2 = 18769$. The digits in their tens' places are 2 and 6,

respectively; both these are even numbers.) This result is easy to prove, and we leave it to the reader to do the needful.

We apply the result thus. Since the last two digits in $\overline{\mathsf{THREE}}$ are both E, it follows that E cannot be odd. Therefore, E is either 4 or 6.

- Since E is even, it follows that $\overline{\mathsf{THREE}}$ is even. So, $\overline{\mathsf{TWO}}$ itself is even, $\therefore$ $\overline{\mathsf{THREE}}$ is divisible by 4.

  We now use the test for divisibility by 4: *In base-10, a number is divisible by 4 iff the 2-digit number formed by its last two digits is divisible by 4.* This implies that the number $\overline{\mathsf{EE}}$ is divisible by 4. Since $\overline{\mathsf{EE}}$ is either 44 or 66, and of these only 44 is divisible by 4, it follows that E $= 4$. This implies, in turn, that O is either 2 or 8.

- The partial reconstruction now has the following appearance:

$$\overline{\mathsf{1WO}}^2 = \overline{\mathsf{1HR44}}.$$

  Since the number on the right is less than 20000, it follows that $\overline{\mathsf{1WO}} \le 141$ (since $\sqrt{20000} \approx 141.42$). As O is 2 or 8, it follows that W < 4; therefore, W is either 2 or 3. (Since T $= 1$, the possibility that W $= 1$ is ruled out.)

- Since O $= 2$ or 8 and W $= 2$ or 3, it follows that $\overline{\mathsf{TWO}}$ is one of the following: 128, 132, 138. (Only three possibilities are left!) The finish is now rather quick: we have, $128^2 = 16384$, $132^2 = 17424$, $138^2 = 19044$. Only the third possibility 'fits' (it is the only one where the last two digits are both 4). So we have our answer:

$$\overline{\mathsf{TWO}} = 138, \qquad \overline{\mathsf{THREE}} = 19044.$$

6.4 Comments

Several comments need to be made concerning the analysis of the two problems in the preceding section.

- In actually solving the problem, much of the working is 'mental'; notation is not used in quite as precise and detailed a manner as that outlined above. This is quite natural. Nevertheless, precise notation does have its place. For one thing, the problem may have multiple solutions, and when this happens, one usually misses out on the alternate solutions if one mentally goes along a linear track. Also, precise notation aids in the analysis itself (as stressed earlier).

- An elaborate record-keeping system was used to keep track of our findings. The comments made above apply here as well.
- The analysis used only elementary mathematical principles, but in a most effective way. This is the case for most cryptarithms.

A simpler example A much more elementary example is the following problem:

$$\begin{array}{r} \mathsf{O}\ \mathsf{N} \\ \times\ \ 3 \\ \hline \mathsf{G}\ \mathsf{O} \\ \hline \end{array}$$

The problem may also be written as $\mathsf{ON}+\mathsf{ON}+\mathsf{ON}=\mathsf{GO}$. Here is how we do the analysis.

- Since $3\times\mathsf{ON}$ is a 2-digit number, $3\times\mathsf{ON}\le 99$, $\therefore\ \mathsf{ON}\le 33$ and $\mathsf{O}\le 3$. Thus, $\mathsf{O}=1$, 2 or 3.
- Since $3\mathsf{N}\equiv\mathsf{O}$, we examine the multiplication table for 3. We deduce the following.

$$\begin{aligned} \mathsf{O}=1 &\quad\Rightarrow\quad \mathsf{N}=7, \\ \mathsf{O}=2 &\quad\Rightarrow\quad \mathsf{N}=4, \\ \mathsf{O}=3 &\quad\Rightarrow\quad \mathsf{N}=1. \end{aligned}$$

- Each of these possibilities gives rise to a solution. The solutions are:

$$17\times 3=51, \quad 24\times 3=72, \quad 31\times 3=93,$$

with $(\mathsf{O},\mathsf{N},\mathsf{G})=(1,7,5)$, $(2,4,7)$ and $(3,1,9)$, respectively. It is clear that this list of solutions is complete.

Note that the requirement about the uniqueness of solution has been violated. As already stated, it takes enormous skill to devise a really good cryptarithm.

Listed below are several cryptarithms. In some cases, multiple solutions exist; your analysis must expose all possibilities. Be forewarned that each problem has a twist of its own; there is no single procedure that will work for all the problems. (Is this not true for any class of problems?)

EXERCISES

6.4.1 Determine all possible solutions (or show that none exist) for the following cryptarithms:

(a) $\overline{\text{ON}} \times 4 = \overline{\text{GO}}$.
(b) $\overline{\text{ON}} \times 5 = \overline{\text{GO}}$.
(c) $\overline{\text{ON}} \times 6 = \overline{\text{GO}}$.

6.4.2 Solve the cryptarithm: $\overline{\text{AH}}^2 = \overline{\text{BAH}}$.

6.4.3 Solve the following cryptarithms (or show that no solutions exist):

(a) $\overline{\text{ABCDE}} \times 4 = \overline{\text{EDCBA}}$.
(b) $\overline{\text{ABCDE}} \times 7 = \overline{\text{EDCBA}}$.
(c) $\overline{\text{ABCDEF}} \times 5 = \overline{\text{FABCDE}}$.

6.4.4 Here are a few particularly elegant cryptarithms, each with a 'double meaning':

(a) $\overline{\text{THREE}} + \overline{\text{THREE}} + \overline{\text{FOUR}} = \overline{\text{ELEVEN}}$.
(b) $\overline{\text{CROSS}} + \overline{\text{ROADS}} = \overline{\text{DANGER}}$.

6.4.5 Solve the following cryptarithms:

(a) $\overline{\text{JUNE}} + \overline{\text{JULY}} = \overline{\text{APRIL}}$.
(b) $\overline{\text{FLY}} + \overline{\text{FOR}} + \overline{\text{YOUR}} = \overline{\text{LIFE}}$.

⋆ ⋆ ⋆

COMMENT. In this account, we have concentrated on cryptarithms of the kind $* + * = *$, where each '$*$' is a word representing some number. There are also problems of the kind $* + * = * + *$. These are typically somewhat harder to solve. Given below is one of the more elegant problems of this variety.

6.4.6 Determine all possible solutions (or show that none exist) for the cryptarithm:

$$\overline{\text{ACID}} + \overline{\text{BASE}} = \overline{\text{SALT}} + \overline{\text{H20}}.$$

Note that in '$\overline{\text{H20}}$', the '0' is a digit, not a letter.

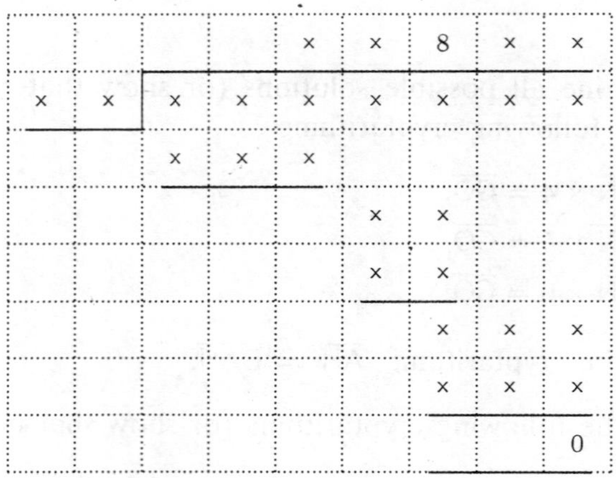

Figure 6.1. *The lonely '8'*

6.5 Faded documents

We now present another type of arithmetical restoration problem, referred to by Degrazia (see list of references) as the 'faded document problem': some digits have 'faded away', and restoration is based on the digits that remain legible. These puzzles are harder to construct than cryptarithms. The requirement of uniqueness is followed as earlier; but this makes them easier to solve than to create! It is usual to use an 'x' in place of the unreadable digits.

EXERCISES

6.5.1 The Lonely '8' Here is a classic—a division problem with a solitary '8'; just one digit is legible (see Figure 6.1)! Rather surprisingly, the restoration is an easy one; indeed, it can be accomplished within 3 to 4 steps at the most.

Hints: (i) The division has only 3 lines, yet the quotient is a 5-digit number; when can this happen? (ii) What are the first and last digits of the quotient? (iii) What is the divisor?

6.5.2 Another Lonely '8' In this problem (Figure 6.2), there are precisely two occurrences of the digit '8', and only one of them is still legible in the document.

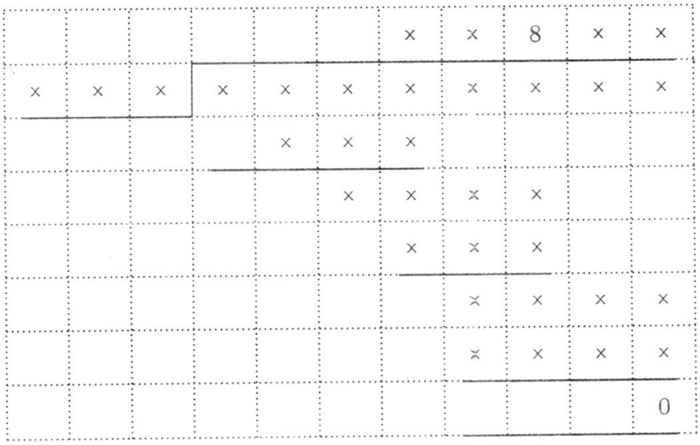

Figure 6.2. *Another lonely '8'*

6.5.3 A Lonely '7' The 7 in this problem (Figure 6.3) is the sole '7' that occurs in the problem and, therefore, has some reason to be lonely. However, the restoration is a fairly easy one.

6.6 A perfect puzzle

Shown below is a string of digits, ⟨P, E, R, F, E, C, T⟩, and its sequence of absolute first differences, ⟨P, U, Z, Z, L, E⟩.

P		E		R		F		E		C		T
	P		U		Z		Z		L		E	

Clearly, we have

$$P = |P - E|, \quad U = |E - R|, \quad Z = |R - F| = |F - E|,$$
$$L = |E - C|, \quad E = |C - T|.$$

The problem is, as earlier, to unravel the code. Note that this problem is of a somewhat different type from the ones we studied earlier. (It has been taken from the puzzle corner of the *New Scientist*.) Is the puzzle really perfect? We shall be able to answer the question only after solving it! We argue as follows:

(a) There are 9 digits used in the cryptarithm, corresponding to the 9 letters P, E, R, F, C, T, U, Z, L. None of these can be 0 (the only possible candidate for 0 is E, because $|P - E| = P$; but $E = 0$

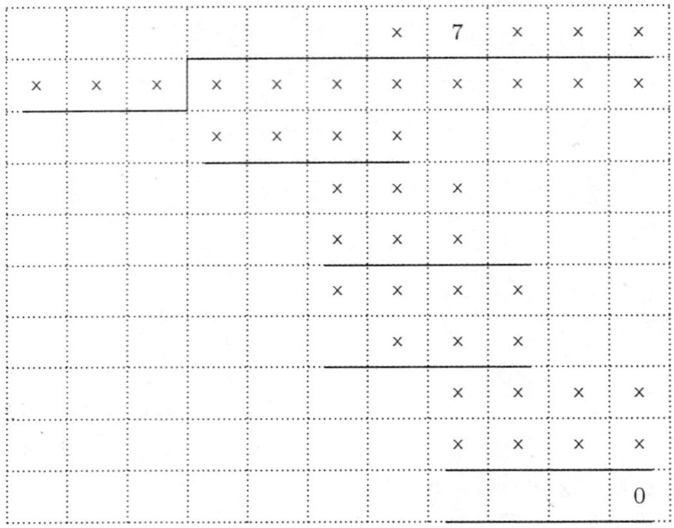

Figure 6.3. *The lonely '7'*

would imply that $\mathsf{R} = \mathsf{U}$, which is not allowed). It follows that

$$\{\mathsf{P}, \mathsf{E}, \mathsf{R}, \mathsf{F}, \mathsf{C}, \mathsf{T}, \mathsf{U}, \mathsf{Z}, \mathsf{L}\}$$

is a permutation of the set $\{1, 2, 3, 4, 5, 6, 7, 8, 9\}$.

(b) Since $\mathsf{E} \neq 0$, we must have $\mathsf{E} = 2\mathsf{P}$ or $\mathsf{P} = \frac{1}{2}\mathsf{E}$. Thus, E is *even*, and therefore E is one of the numbers 2, 4, 6 or 8.

(c) Next, $|\mathsf{R} - \mathsf{F}| = \mathsf{Z} = |\mathsf{F} - \mathsf{E}|$. Since $\mathsf{E} \neq \mathsf{R}$, this implies that

$$\mathsf{R} - \mathsf{F} = \mathsf{F} - \mathsf{E}, \quad \text{i.e.,} \quad \mathsf{R} + \mathsf{E} = 2\mathsf{F}.$$

Therefore, $\mathsf{R} = 2\mathsf{F} - \mathsf{E} = 2(\mathsf{F} - \mathsf{P})$, and R is also even.

(d) Working from the right side, we get: $|\mathsf{E} - \mathsf{C}| = \mathsf{L}$, so $\mathsf{E} - \mathsf{C} = \pm\mathsf{L}$ or $\mathsf{C} = \mathsf{E} \pm \mathsf{L}$; and $|\mathsf{T} - \mathsf{C}| = \mathsf{E}$, so $\mathsf{T} - \mathsf{C} = \pm\mathsf{E}$ and $\mathsf{T} = \mathsf{C} \pm \mathsf{E}$.

Therefore, $\mathsf{T} = (\mathsf{E} \pm \mathsf{L}) \pm \mathsf{E}$ for some combination of $\pm$ signs. However, some combinations are quickly ruled out: $\mathsf{T} = \mathsf{E} + \mathsf{L} - \mathsf{E}$ would give $\mathsf{T} = \mathsf{L}$, which is illegal, and $\mathsf{T} = \mathsf{E} - \mathsf{L} - \mathsf{E}$ would make either T or L negative, which too is illegal. Thus, the second sign must be '+' and we must have

$$\mathsf{T} = \mathsf{C} + \mathsf{E}, \qquad \mathsf{L} = \pm(\mathsf{E} - \mathsf{C}).$$

So we have: $\mathsf{P} = \frac{1}{2}\mathsf{E}$, $\mathsf{R} = 2\mathsf{F} - \mathsf{E}$, $\mathsf{T} = \mathsf{C} + \mathsf{E}$, $\mathsf{U} = 2\,|\mathsf{F} - \mathsf{E}|$, $\mathsf{Z} = |\mathsf{F} - \mathsf{E}|$ and $\mathsf{L} = |\mathsf{C} - \mathsf{E}|$. (We have expressed all the quantities in terms of

E, F and C.) The problem may now be stated succinctly as follows: assign (digital) values to E, F and C so that the numbers

$$\frac{E}{2}, E, F, 2F - E, 2|F - E|, |F - E|, C, C + E, |C - E|$$

are the same as 1, 2, 3, 4, 5, 6, 7, 8, 9, in some order.

(e) The fact that the sum $1 + 2 + 3 + \cdots + 8 + 9 = 45$ is a multiple of 3 now comes to our aid, for it implies that the sum

$$\tfrac{1}{2}E + E + F + 2F - E + 2|F - E| + |F - E| + C + C + E + |C - E|$$

is a multiple of 3; equivalently, that

$$\frac{3}{2}E + 3F + 3|F - E| + 2C + |C - E|$$

is a multiple of 3. This implies, in turn, that $2C + |C - E|$ is a multiple of 3, so we deduce that:

- if $C > E$, then $2C + C - E$ is a multiple of 3, so E is a multiple of 3 (implying that $E = 6$, since E is known to be even); and
- if $C < E$, then $2C + E - C$ is a multiple of 3, i.e., $C + E$ is a multiple of 3.

(f) The first possibility listed above cannot occur, for it implies that $C + E \geq 13$ (because $E = 6$ and $C > E$), but we know that $C + E = T$ is a digit. It follows that $C < E$ and, therefore, that $C + E$ is a multiple of 3.

(g) We shall now try out the various possibilities.

- $E = 2$ leads to $C = 1$ (from (f)) and $P = \frac{1}{2}E = 1$, which has a duplication of digits and so is illegal.
- $E = 4$ leads to $C = 2$ (because $C + E$ is a multiple of 3) and $P = \frac{1}{2}E = 2 = C$, which is illegal too.
- $E = 6$ leads to the same difficulty.
- So, $E = 8$ (the only possibility left—it had better work!), $C = 1$, $P = 4$, $T = C + E = 9$, $L = |E - C| = 7$.

  Next, $2F > E$, so $F > 4$; $\therefore$ $F = 5$ or 6 (the values 7, 8 and 9 having already been blocked by L, E and T, respectively). The possibility $F = 6$ leads to $Z = |F - E| = 4 = P$, $\therefore$ a duplication of digits, so $F = 5$, and now everything falls into place.

$$\mathsf{R} = 2\mathsf{F} - \mathsf{E} = 2,$$
$$\mathsf{U} = 2\,|\mathsf{F} - \mathsf{E}| = 6.$$

Therefore, we have $\mathsf{P} = 4$, $\mathsf{E} = 8$, $\mathsf{R} = 2$, $\mathsf{F} = 5$, $\mathsf{C} = 1$, $\mathsf{T} = 9$, $\mathsf{U} = 6$, $\mathsf{Z} = 3$ and $\mathsf{L} = 7$.

⋆ ⋆ ⋆

The puzzle has been solved. Note that the answer is unique—the puzzle is indeed perfect! As remarked earlier, it requires great skill to create such a puzzle. Note how the test for divisibility by 3 simplified our search.

6.7 The game of Bulls & Cows

The game featured here, *Bulls & Cows*, is well-known, but it appears under different names and it may take the reader some time to recognize it and find that it is merely an old friend in disguise.[2]

Rules of the game The game is played between two players, ***A*** and ***B***, as follows. ***A*** selects a 4-digit number N_A with distinct digits, and ***B*** tries to find N_A by making various guesses, N_B, each guess being a 4-digit number. After each guess, ***A*** provides feedback to ***B*** on how 'close' N_B is to N_A. The manner in which this is done is described below.

If N_A and N_B share a digit in the same place, ***A*** announces '1 bull', and if there are x such digits, then the announcement is 'x bulls'. (If N_A and N_B are identical then ***A*** responds with '4 bulls' and this ends the game.) If N_A and N_B have digits in common that are not in the same place, then ***A*** announces '1 cow', '2 cows', etc., as the case may be. Thus, 'x bulls, y cows' would indicate that there are x digits guessed correctly and placed correctly, and y digits guessed correctly but placed incorrectly; '4 cows' would indicate that N_A and N_B are derangements of one another (that is, having the same digits but completely 'out-of-sync' with one another).

To illustrate, suppose that $N_A = 3721$. For various guesses by ***B***, we give the response from ***A*** ('B' stands for 'Bull', 'C' for 'Cow').

[2]Two well-known games related to *Bulls & Cows* are *Mastermind* and *Destroyer*.

Guess (N_B)	3719	1706	1111	2345	2367	1237
Response	2B, 1C	1B, 1C	1B	2C	3C	4C

The pattern should be clear. Obviously, ***B***'s task is to find N_A, using as few guesses as possible.

Several variations are possible on the basic model given above.

- We could work with 5-digit or 6-digit numbers to increase the complexity and level of difficulty of the game.
- We could allow repetition of digits in N_A.
- We could play with *words* instead of numbers, and (to complicate matters) insist that the selection and guesses all be dictionary words. Since there are more letters than digits, the level of difficulty here is correspondingly greater, especially with the constraint that only dictionary words be used.

We shall study only the simplest model: 4-digit numbers, with no repetition of digits. The problems will be given in the following format: the guesses (N_B) will be listed with the corresponding responses provided alongside. The reader's task will be to deduce the selected number N_A from the given responses.

Here is a worked-out example. We are asked to find N_A, if the responses by ***A*** to guesses from ***B*** are as follows:

Guess (N_B)	1209	3012	9218	8910
Response	3C	1B, 1C	1B, 2C	4C

For convenience, we write N_A as $\overline{\mathsf{abcd}}$, with d as the units digit, c as the tens digit, and so on. The last response tells us that N_A is a derangement of 8910. Combining this deduction with the other responses, we deduce the following:

$$\mathsf{a} = 0 \text{ or } 9, \qquad \mathsf{b} = 0,\ 1 \text{ or } 8,$$
$$\mathsf{c} = 8 \text{ or } 9, \qquad \mathsf{d} = 1 \text{ or } 8.$$

In the response '1 bull, 1 cow' to the guess 3012, the 'bull' must be 0 and not 1 (because c = 8 or 9); thus, b = 0. This forces a to be 9, and, in turn, we deduce that c = 8, d = 1. Therefore, $N_A = 9081$, and we have the required answer.

This is an actual game score. Note that ***B*** was rather lucky to have made a good first guess.

We shall leave open the question of what is the best strategy to follow in this game and, with this strategy, what is the largest number of guesses required to find the number (in the 'worst-possible' scenario).

⋆ ⋆ ⋆

In the problems listed below, the response to each guess is given in brackets, using the abbreviation 'B' for bull and 'C' for cow. (The problems are not as easy as the one solved above!)

EXERCISES

6.7.1 3824 (3C), 6348 (2C), 2489 (1B, 1C), 2354 (4C).

6.7.2 8264 (3C), 4026 (1B, 1C), 4692 (2B), 4802 (1B, 1C).

6.7.3 2463 (1B, 1C), 4573 (3C), 3458 (2C), 2735 (1B, 1C).

6.7.4 5634 (2C), 2341 (1B, 1C), 2560(1C), 7386 (1B, 1C), 7845 (1B, 1C), 9375 (1C), 8946 (2B).

6.7.5 7643 (1C), 1257 (2C), 9875 (2C), 8720 (1B), 8561 (2C), 5190 (1B, 1C).

6.8 References

[1] Ball, W W Rouse. *Mathematical Recreations and Essays*, Macmillan, 1960.

[2] Brooke, Rosey. *Cryptarithmetic*, Dover.

Chapter 7

Magic Squares

7.1 Introduction

Magic squares constitute a topic with a long and distinguished past; indeed, the name itself ('magic square') should tell us so. Nowadays, we think of them essentially as arrays of numbers with numerous appealing arithmetical properties and symmetries, but in earlier eras they have been associated with astrology, magic, alchemy, and other such exotica. In 2000 BC or so, a square of order 3, called the *Lo Shu*, was the symbol of the Chinese empire. In his famous painting *Melancholia*, the painter Albrecht Dürer included a fourth-order magic square—the year in which the painting was done (1514) features in the square. (A reproduction of the painting is given in Martin Gardner's book [1].) In this chapter, however, we eschew such complications and concentrate only on the arithmetical aspects of magic squares.

Definition of magic square A magic square of order n consists of a $n \times n$ array of numbers, typically $\{1, 2, 3, \ldots, n^2\}$, arranged so that the sums of the entries in the n rows, the n columns and the two main diagonals are all equal. The common sum is termed the *magic sum*, and it is clearly equal to

$$\frac{1 + 2 + 3 + \cdots + n^2}{n} = \frac{n(n^2 + 1)}{2}.$$

The table given below displays the values of the magic sum for a few values of n.

Order of square	3	4	5	6	7
Magic sum	15	34	65	111	175

The following is an example of a 4×4 magic square, with the numbers 1, 2, 3, ..., 16 as its entries.

1	8	13	12
14	11	2	7
4	5	16	9
15	10	3	6

The magic sum is 34, in agreement with the values listed above.

There are many obvious questions we may want to ask about magic squares.

EXAMPLE. Using the base set $\{1, 2, 3, \ldots, n^2\}$, how many $n \times n$ magic squares exist? What if we consider other sets of numbers? Are there magic squares whose entries are all squares, or all cubes, or all primes? And so on, ...; the list is easily extended. Note that there are two ground rules that we shall generally not relax during our enquiry:

- Repetitions of numbers are not permitted—the numbers used in a magic square must be distinct.
- The numbers used must be positive integers.

Also, quite reasonably, we regard two magic squares as the same ('congruent' is a better word) if, by using reflections and rotations, one square can be transformed into the other. Thus, the three 4×4 magic squares shown below are regarded as essentially the same.

7	12	1	14
2	13	8	11
16	3	10	5
9	6	15	4

7	2	16	9
12	13	3	6
1	8	10	15
14	11	5	4

9	16	2	7
6	3	13	12
15	10	8	1
4	5	11	14

COMMENT. The magic squares shown above have many properties beyond those required by the definition of a magic square. Martin Gardner calls them *diabolic*, and provides the noteworthy information that the left-most square, shown is from Khajuraho in India. A magic square is called diabolic if in addition to its row-, column- and diagonal-sums being equal (to the magic sum), the 'broken diagonals' too have this sum. Here, a broken diagonal is one that breaks off at the edge of the square, and reappears at the corresponding place on the opposite edge. In the left-most square shown above, the examples of broken diagonals are $(2, 12, 5, 15)$, $(16, 13, 1, 4)$ and $(1, 11, 6, 16)$. Note that each has a sum of 34. An additional property that these squares have is that any 2×2 block of contiguous squares has a sum of 34; e.g., the block containing the numbers 8, 11, 5 and 10. 'Diabolic' is the right word to use!

7.2 Squares of order 3

Trivially, any 1×1 square is magic, and no 2×2 square can be magic unless all its entries are the same, so we commence our survey with an inquiry into third order squares. Our first question is: *Using the set* $\{1, 2, \ldots, 9\}$, *how many essentially different* 3×3 *magic squares are there?*

Write the square as:

a	b	c
d	e	f
g	h	i

The magic sum is $(1+2+\cdots+8+9)/3 = 45/3 = 15$. This implies that

$$a+b+c = d+e+f = g+h+i = 15,$$
$$a+d+g = b+e+h = c+f+i = 15,$$
$$a+e+i = c+e+g = 15.$$

Consider the four sums containing e (the sums corresponding to the diagonals and the middle row and middle column). Adding the four sums to one another, we find that each letter occurs just once in the sum, except for e which occurs four times. It follows that

$$(a+b+c+d+e+f+g+h+i) + 3e = 4 \times 15 = 60.$$

The bracketed sum is 45, for the numbers occurring within are 1, 2, 3, ..., 9. Therefore, $45 + 3e = 60$ and so $e = 5$. Thus, the central entry is 5, and we have found a property common to all 3×3 magic squares based on the set $\{1, 2, 3, \ldots, 9\}$: *the central entry is 5.*

Consider next the location of the '1'. By rotation or reflection, we can ensure that the '1' occurs either in the top left-hand corner or in the middle of the top row. (These operations do not affect the central entry.) This means that we need to examine only the possibilities $a = 1$ and $b = 1$, respectively.

Suppose $a = 1$. Since the magic sum is 15 and $e = 5$, we see that $i = 9$. Here is the incomplete square that we have till now:

1	b	c
d	5	f
g	h	9

Consider the different ways in which 1 can be combined with two other unequal numbers to yield a sum of 15. There are just two ways: $1 + 5 + 9$ and $1 + 6 + 8$. We have already used $1 + 5 + 9$ on the diagonal, and we require two more ways—one each for a row and a column. This is clearly not possible. So, with $a = 1$ we cannot complete the magic square. Therefore, *the '1' cannot occupy a corner square.*

Now, consider now the possibility $b = 1$, which forces $h = 9$:

a	1	c
d	5	f
g	9	i

Examining the top row, we deduce that $(a, c) = (6, 8)$ or $(8, 6)$. Suppose that $a = 6$ and $c = 8$. Since the magic sum is 15, the rest of the entries can be filled in immediately; since $a + 5 = 11$, we have $i = 4$, and since $i + 9 = 13$, we have $g = 2$; so, we get

6	1	8
7	5	3
2	9	4

A simple check reveals that the required sums are all 15. We have constructed our first magic square!

With $a = 8$ and $c = 6$, we obtain

8	1	6
3	5	7
4	9	2

This is merely a reflected version of the earlier square, which is not surprising as the intermediate configurations:

6	1	8
	5	
	9	

8	1	6
	5	
	9	

are themselves reflected versions of one another. So, we have a noteworthy result: *Using the numbers 1, 2, 3, 4, 5, 6, 7, 8, 9, essentially only one* 3×3 *magic square can be made.*

7.3 Problems

7.3.1 Investigate the problem of constructing a 3×3 magic square using the set

$$\{11, 12, 13, 21, 22, 23, 31, 32, 33\}.$$

How is this related to the problem of constructing a magic square using the set $\{1, 2, 3, \ldots, 8, 9\}$?

7.3.2 The following square has been partially filled in. Fill in the missing entries and convert it into a magic square. There are no restrictions on the number set to be used (apart from the usual ones—only positive integers to be used, and no repetition of numbers).

3	17	
	7	

7.3.3 Repeat this for the incomplete square shown below.

14	11	
		19

7.3.4 Let $a_1, a_2, \ldots, a_8, a_9$ be nine positive integers in AP.[1] We wish to construct a 3×3 magic square using these nine numbers. Show that a_5 must occupy the central square.

[1] A sequence of numbers $a_1, a_2, a_3, a_4, \ldots$ is said to be in AP, or in "arithmetic progression", if $a_1 - a_2 = a_2 - a_3 = a_3 - a_4 = \cdots$. E.g., the sequence 1, 3, 5, 7, 9, ... is in AP, as is 4, 9, 14, 19, 24, 29, ...

7.4 Squares of order 4

We now move to the study of 4×4 squares. Unlike the 3×3 case, no claim can be made here about uniqueness; indeed, we have already seen two essentially different 4×4 magic squares earlier in this chapter (Section 7.1).

However, many striking properties still hold, and these can be used to construct new 4×4 magic squares. As the reader will see shortly, if the number set is not fixed in advance, it is rather simple to construct such squares.

With the increased number of entries, it proves more convenient to use subscripts, and we shall use the model shown below.

a_1	a_2	a_3	a_4
b_1	b_2	b_3	b_4
c_1	c_2	c_3	c_4
d_1	d_2	d_3	d_4

EXERCISES

7.4.1 Assuming that the 4×4 square shown above is a magic square, prove the following equalities:

$$a_1 + a_4 = d_2 + d_3,$$
$$a_2 + a_3 = d_1 + d_4,$$
$$b_1 + b_4 = c_2 + c_3,$$
$$b_2 + b_3 = c_1 + c_4.$$

7.4.2 Prove the following:

$$a_1 + d_4 = b_3 + c_2,$$
$$d_1 + a_4 = b_2 + c_3,$$

and list further relations of this kind.

★ ★ ★

The symmetries and complementarity properties indicated above are quite striking, and they allow for a simple construction of a 4×4 magic square for which one of the rows has been fixed in advance. To illustrate how, let the first row be $(22, 12, 18, 87)$. (This corresponds to the date of birth of the great Indian mathematician

Srinivasa Ramanujan: 22nd December 1887.) The magic sum for the square is now fixed: $22 + 12 + 18 + 87 = 139$. We start by assigning values to d_1, d_4, c_2, b_3, c_3 and b_2 in turn, in the following manner.

22	12	18	87
	b_2	b_3	
	c_2	c_3	
d_1			d_4

This is how we argue the matter.

- Since $a_2 + a_3 = 12 + 18 = 30$, we must have $d_1 + d_4 = 30$. We select any two positive integers, distinct from 12 and 18 (which have already been 'used'), whose sum is 30.

  COMMENT. At this and many subsequent points, numerous choices will be available to us. The general rule is: *any choice will do, provided that it avoids repetition of numbers.* In some cases we may get 'stuck' at a later point, i.e., we may be left with no choices. In such cases, we must backtrack and choose afresh.

- We arbitrarily take $d_1 = 1$ and $d_4 = 29$ as the two numbers. We now have:

22	12	18	87
	b_2	b_3	
	c_2	c_3	
1			29

  Here, $a_1 + d_4 = 22 + 29 = 51$; $\therefore\ c_2 + b_3 = 51$. We choose suitable values for c_2, b_3, say: $c_2 = 11$, $b_3 = 40$. We now obtain:

22	12	18	87
	b_2	40	
	11	c_3	
1			29

  Next, we must have $b_2 + c_3 = 1 + 87 = 88$, and a possible choice is $b_2 = 57$, $c_3 = 31$:

22	12	18	87
	57	40	
	11	31	
1	d_2	d_3	29

Since the magic sum is 139, the entries d_2 and d_3 emerge on their own:

$$d_2 = 139 - (2 + 57 + 11) = 139 - 80 = 59,$$
$$d_3 = 139 - (18 + 40 + 31) = 139 - 89 = 50,$$

and we obtain the following updated picture:

22	12	18	87
b_1	57	40	b_4
c_1	11	31	c_4
1	59	50	29

Let $b_1 = x$; then $c_1 = 116 - x$, $b_4 = 42 - x$, and $c_4 = x - 19$. The square now has the following appearance:

22	12	18	87
x	57	40	$42 - x$
$116 - x$	11	31	$x - 19$
1	59	50	29

We select x so that the numbers x, $166 - x$, $42 - x$ and $x - 19$ are positive and distinct from one another as well as the numbers already used, namely:

$$22,\ 12,\ 18,\ 87,\ 57,\ 40,\ 11,\ 31,\ 1,\ 59,\ 50.$$

The requirement of positivity implies that

$$20 \leq x \leq 41.$$

The trial values $x = 20$, 21 and 22 fail because of repetitions (please check!), but $x = 23$ works, and we do obtain a 4×4 magic square:

22	12	18	87
23	57	40	19
93	11	31	4
1	59	50	29

(Check that the required totals do work out to 139.)

In view of the numerous choices available at practically every step, it seems clear that the number of magic squares whose top row has the numbers $(22, 12, 18, 87)$ must be fairly large.

EXERCISES

7.4.3 Construct a magic square with $(15, 8, 19, 47)$ as its first row. (This corresponds to India's date of Independence: 15th August 1947, or 15–8–1947.)

7.4.4 Construct a magic square with $(23, 6, 19, 93)$ as its first row. (This corresponds to the day on which Andrew Wiles announced that he had proved the notorious conjecture known as 'Fermat's Last Theorem'. As fate would have it, the proof as announced had a flaw, but Wiles managed to 'fix' the proof over the course of the following year.)

COMMENT. This technique does not tell us how to construct a 4×4 magic square for which the set of entries has been fixed in advance. Indeed, it is not clear how one would fix even the first row in such an instance.

Ramanujan's contribution It is of interest to include here a quote on the subject of magic squares from the notebooks of Srinivasa Ramanujan, who evidently studied the topic very seriously when he was a young boy.

> ... Magic squares can be constructed by combining two sets of letters so that the same letter may not appear [twice] in a row, a column [or a diagonal]; for example, if we want to construct a square containing n rows and n columns, we should take two sets of n letters A, B, C, D, E, ..., and P, Q, R, S, T, ..., combine the two as $A+P$, $A+Q$, $A+R$, ..., $B+P$, $B+Q$, $B+R$, ..., $C+P$, $C+Q$, $C+R$, ..., and arrange them in such a way that a given letter may not appear [twice] in the same row, column [or diagonal] ....

Ramanujan goes on to point out that:

- If $A+P$, $A+Q$, $A+R$, ..., are in AP ('arithmetic progression'), then $B+P$, $B+Q$, $B+R$, ..., too are in AP.
- If $A+P$, $A+Q$, $A+R$, $A+S$, ... and $B+R$ are known, then $B+P$, $B+Q$, $B+S$, ..., too are known.
- It is more convenient to assign values to $A+P$, $A+Q$, ..., than to assign values separately to A, B, C, ..., and to P, Q, R, ...

COMMENT. The quote needs some explanation. Considerable work has been done on the design of square arrays of letters that have the property described by Ramanujan—the property that each of the rows and columns contains the same set of letters, arranged so that there is no repetition of letters in any row or column. For $n = 3$, two such arrays are shown here.

C	A	B
A	B	C
B	C	A

Q	P	R
R	Q	P
P	R	Q

(Note that the array at the right is a reflected version of one given earlier, with the substitutions $A \to P$, $B \to Q$, $C \to R$.) We now combine the two arrays by simple addition, as shown below.

$C+Q$	$A+P$	$B+R$
$A+R$	$B+Q$	$C+P$
$B+P$	$C+R$	$A+Q$

The row and column sums of this array are $A+B+C+P+Q+R$, and the diagonal sums are $A+B+C+3Q$ and $P+Q+R+3B$, respectively. If we now require that (A, B, C) and (P, Q, R) be in AP, then $A+B+C = 3B$ and $P+Q+R = 3Q$, and now all the relevant sums are equal.

EXAMPLE. Take $(A, B, C) = (10, 20, 30)$ and $(P, Q, R) = (1, 2, 3)$; we then obtain the array shown below.

32	11	23
13	22	31
21	33	12

This is a magic square with a magic sum of 66. With $(A, B, C) = (0, 3, 6)$ and $(P, Q, R) = (1, 2, 3)$, we obtain a rotated version of an old friend, namely:

8	1	6
3	5	7
4	9	2

The idea should now be clear: we must find two distinct arrays of letters of the type described by Ramanujan, and the two arrays must 'fit' with one another in the right manner. Can this idea be used for the 4×4 square? Keeping in mind the equalities mentioned earlier, which hold for all 4×4 magic squares, we start with the arrays shown below.

A	D	C	B
C	B	A	D
B	C	D	A
D	A	B	C

P	S	Q	R
R	Q	S	P
S	P	R	Q
Q	R	P	S

Note that the relations listed earlier hold: $a_1 + a_4 = d_2 + d_3, \ldots,$ and that the second array has a different structure from the first one. By addition, we obtain the following array.

$A+P$	$D+S$	$C+Q$	$B+R$
$C+R$	$B+Q$	$A+S$	$D+P$
$B+S$	$C+P$	$D+R$	$A+Q$
$D+Q$	$A+R$	$B+P$	$C+S$

(The phrase 'fit together' will be clearer now: in this array, no sum has been repeated; the sixteen combinations $A+P$, $D+S$, $C+Q$, $B+R$, ... are all distinct.)

With $(A, B, C, D) = (0, 4, 8, 12)$ and $(P, Q, R, S) = (1, 2, 3, 4)$ we get the array shown below.

1	16	10	7
11	6	4	13
8	9	15	2
14	3	5	12

This array is about as magical as one may wish, for it has many properties in addition to the magic square property. (However, it is not diabolic.) The reader should examine the square and list these additional features. Note that the condition that (A, B, C, D) and (P, Q, R, S) be in AP is not needed here. For example, the choice $(A, B, C, D) = (0, 4, 8, 12)$ and $(P, Q, R, S) = (3, 1, 4, 2)$ yields the following magic square.

3	14	9	8
12	5	2	15
6	11	16	1
13	4	7	10

This square is wholly different! Clearly, other magic squares can be generated using the same basic structure, merely by taking different permutations of $(0, 4, 8, 12)$ for (A, B, C, D) and $(1, 2, 3, 4)$ for (P, Q, R, S). We cannot, of course, count these magic squares as distinct—this would amount to over-counting, as the effects of reflection and rotation have not been considered.

Other such families of 4×4 magic squares can be found, e.g., by using the arrays shown below.

A	D	D	A
B	C	C	B
C	B	B	C
D	A	A	D

P	Q	R	S
S	R	Q	P
S	R	Q	P
P	Q	R	S

The following restrictions are needed:

$$A + D = B + C, \quad P + S = Q + R.$$

(These are necessary for the row sums to work out right in the $ABCD$ array and for the column sums to work out right in the $PQRS$ array.) The magic square resulting from this is shown below.

$A+P$	$D+Q$	$D+R$	$A+S$
$B+S$	$C+R$	$C+Q$	$B+P$
$C+S$	$B+R$	$B+Q$	$C+P$
$D+P$	$A+Q$	$A+R$	$D+S$

If we take $(A, B, C, D) = (0, 4, 8, 12)$ and $(P, Q, R, S) = (1, 3, 2, 4)$ (note that the conditions $A + D = B + C$, $P + S = Q + R$ both hold), we obtain the following square.

1	15	14	4
8	10	11	5
12	6	7	9
13	3	2	16

This magic square has some new features of interest, different from those seen earlier. For example, by 'flipping' the two main diagonals (that is, turning each diagonal about the centre of the square through an angle of 180°), we obtain the square below.

16	15	14	13
8	7	6	5
12	11	10	9
4	3	2	1

This certainly comes as a surprise, and it suggests an additional technique for making a 4×4 magic square, though it is difficult to see how one could have thought of it in the first place.

⋆ ⋆ ⋆

COMMENT. It is not clear whether the above technique generalizes in any obvious manner; for instance, can similar statements be made for magic squares of orders 3, or 5, or 6? It seems unclear at this point whether the observation can lead to a technique. Polya once remarked that a method that works just once is a trick and not a technique; it cannot be considered as part of mathematics! Does the aphorism apply here? (Please see Problems 7.5.7.)

⋆ ⋆ ⋆

The field at this stage is wide open! There are numerous questions that suggest themselves and one may quickly get overwhelmed by their number. Here are a few such leads.

(i) Using the "$\{A, B, C, D\}$—$\{P, Q, R, S\}$" technique, can we produce every possible 4×4 magic square based on the number set $\{1, 2, \ldots, 15, 16\}$? (This seems doubtful, but how would one investigate this?)

(ii) How many 4×4 magic squares are there based on the number set $\{1, 2, \ldots, 15, 16\}$? (Squares obtainable from one another by rotations or reflections are not counted as distinct. Obviously, each given magic square has seven others obtainable from it using these operations.)

(iii) How would this analysis extend to the construction of $n \times n$ magic squares for $n > 4$?

7.5 Problems

7.5.1 Prove that, using the array displayed below,

$A+P$	$D+S$	$C+Q$	$B+R$
$C+R$	$B+Q$	$A+S$	$D+P$
$B+S$	$C+P$	$D+R$	$A+Q$
$D+Q$	$A+R$	$B+P$	$C+S$

we can generate 72 distinct magic squares with the number set $\{1, 2, \ldots, 15, 16\}$. Here, 'distinct' means that no two of the squares are obtainable from one other by a sequence of rotations and reflections. (*Hint.* $72 = 24 \times 24/8$.)

7.5.2 Prove, likewise, that using the array displayed below,

$A+P$	$D+Q$	$D+R$	$A+S$
$B+S$	$C+R$	$C+Q$	$B+P$
$C+S$	$B+R$	$B+Q$	$C+P$
$D+P$	$A+Q$	$A+R$	$D+S$

(here, the relations $A+D=B+C$, $P+S=Q+R$ must hold) we can generate 8 distinct magic squares based on the number set $\{1, 2, \ldots, 15, 16\}$ ('distinct' as defined above).

7.5.3 Examine the possibility of constructing a 6×6 magic square with the set $\{1, 2, \ldots, 35, 36\}$, via the following procedure. Group the numbers as follows:

$$\begin{aligned} A &= \{1, 2, \ldots, 9\}, & \text{sum of entries } &= 45 \\ B &= \{10, 11, \ldots, 18\}, & \text{sum of entries } &= 126 \\ C &= \{19, 20, \ldots, 27\}, & \text{sum of entries } &= 207 \\ D &= \{28, 29, \ldots, 36\}, & \text{sum of entries } &= 288, \end{aligned}$$

and construct 3×3 magic squares using each of the sets A, B, C, D. Arrange these four squares as follows.

A	C
D	B

We now have a 6×6 array which is not yet a magic square, though its column sums are equal and its first three rows share the same sum, as do its last three rows. We now check the feasibility of interchanging certain pairs of numbers in A and D, so as to obtain a 6×6 magic square. For simplicity, assume that the four 3×3 magic squares are of the 'same shape'; that is, obtainable from one another by addition of a fixed number to each entry. Symbolically, one may write $B = A + 9$, $C = B + 9$, $D = C + 9$, and the array as:

A	$A + 18$
$A + 27$	$A + 9$

7.5.4 Do magic rectangles exist? (A magic rectangle is a rectangular array of numbers whose row and column totals are all the same. The term 'rectangle' of course includes 'square', but here we explicitly consider the non-square case, when the length exceeds the breadth. Note that it makes no sense to talk of a diagonal sum for such a rectangle.)

Here is an example of a 2×3 magic rectangle:

1	2	−3
−1	−2	3

Its magic sum is 0. Note that we have used negative numbers in the array.

(a) Prove that the magic sum of a non-square magic rectangle is 0. (It follows from this that the use of negative numbers cannot be avoided.)

(b) Prove that every 2×3 magic rectangle is of the following variety, for some choice of a and b:

a	b	$-a-b$
$-a$	$-b$	$a+b$

7.5.5 The above problem suggests that non-square magic rectangles are not particularly worthy of deep study and ought not to be called magical! Can they be made more magical? Here is one possibility: we shall call a rectangular array of numbers 'magic' *if its row-averages and column-averages are all the same*; i.e., if the averages of the numbers in each row and column are the same.

EXAMPLE. The array displayed below

2	3	7
6	5	1

is magical in this sense and has a 'magic mean' of 4.

This is certainly a more satisfactory definition of a magic rectangle. (Of necessity, there is no mention of diagonals.)

Problem Find all 2×3 magic rectangles ('magic' in the sense of averages).

COMMENT. The following is known to be true, though non-trivial to prove: *If $n \geq 4$ is an even integer, then there exists a magic rectangle of dimensions $2 \times n$, with entries from the list 1, 2, 3, ..., $2n-1$, $2n$.* (As earlier, 'magic' is used in the sense of averages.) Note that the magic mean is $(2n+1)/2$. Here is an example with $n = 4$:

1	7	6	4
8	2	3	5

And here is an example with $n = 6$:

1	11	3	9	8	7
12	2	10	4	5	6

The magic means are $4\frac{1}{2}$ and $6\frac{1}{2}$, respectively.

7.5.6 It has been claimed ([1], Chapter 12, page 104) that with the set $\{1, 2, \ldots, 15, 16\}$, a total of 880 distinct magic squares exist ('distinct' used as earlier). How would you verify such a claim?

COMMENT. The number of 5×5 magic squares is not known.

7.5.7 Investigate the possibility of constructing an 8×8 magic square via the idea of 'diagonal flips'. (The array is viewed as being composed of four 4×4 arrays, on each of which suitable flips are effected, following which exchanges are made between the four arrays.)

7.6 Construction of odd-order squares

The knight move. Many students discover for themselves the following method for the construction of a 3×3 magic square based on the list 1, 2, ..., 8, 9. Start with the middle cell of the first column, and place a '1' in it. Now make a *knight move* (1 step up, 2 steps to the right) and place the next integer (in this case, 2) in the cell reached. (The array is considered to wrap around upon itself, top edge upon bottom edge and right edge upon left edge, creating a doughnut-shaped tube, so if a move carries you outside the square then we know right away where we are actually supposed to move.) Once again make a knight move, and place the next integer in the cell reached. If the cell reached happens to be already occupied then, instead of a knight move, *move one step to the right.* Continue in this manner till all the integers till 9 are 'used up'. The resulting array will be found to be the familiar 3×3 magic square:

6	7	2
1	5	9
8	3	4

Observe carefully the moves from 1 to 2 and from 2 till 3; they are both knight moves (1 step up, 2 steps to the right)—with wrap-around. The move from 3 to 4 simply consists of a unit move to the right. The moves from 4 to 5 and from 5 to 6 are knight moves (with wrap-around, as earlier), and the move from 6 to 7 is a unit move to the right. The moves from 7 to 8 and from 8 to 9 are wrap-around knight moves once again.

What is remarkable about this method is that it allows for a vast generalization. Firstly, we may apply the same idea, verbatim, to get a 5×5 magic square. The following array is obtained.

9	15	16	22	3
20	21	2	8	14
1	7	13	19	25
12	18	24	5	6
23	4	10	11	17

It may be verified that this is a magic square. The same idea works for a 7×7 square, and indeed for any odd-order square.

Second, the knight move used has nothing particularly special about it; numerous variations are possible on the same theme. We shall describe below another such scheme.

de la Loubère's method Let $n = 2m+1$; we show how to construct a $n \times n$ magic square by specifying where each of the numbers $1, 2, 3, \ldots, n^2$ is to be placed. We move from cell to cell in a definite manner and number the cells consecutively. Here are the details. Let S_n refer to the set $\{0, 1, 2, \ldots, n-1\}$.

- Number the rows and columns from 0 to $n-1$ as shown below, and place the number 1 in the middle of the top row; i.e., in cell $(m, 2m)$. (In the figure we have taken $n = 5$ for convenience.)

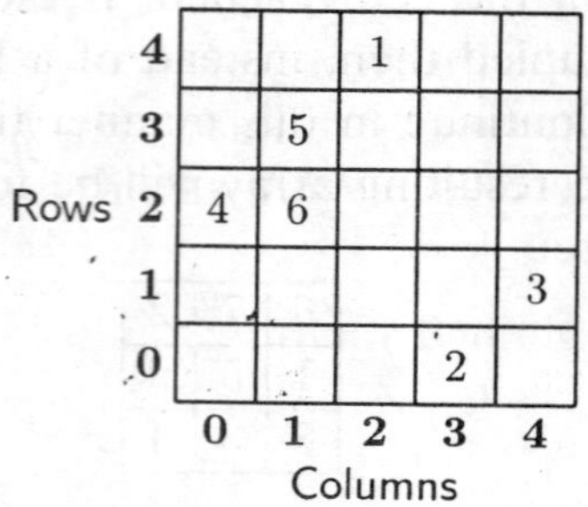

- Suppose that we have just placed the number k in cell (x, y); then we place the number $k+1$ in cell (x', y'), where

$$(x', y') = \begin{cases} (x+1, y+1) \pmod{n} & \text{if } k \text{ is not a multiple of } n, \\ (x, y-1) \pmod{n} & \text{if } k \text{ is a multiple of } n. \end{cases}$$

The $\pmod{n}$ accomplishes the wrap-around referred to earlier.

In the figure above, we have shown how the numbers from 1 to 6 are placed. Observe that at each stage the cell coordinates change by either $(1,1)$ or $(0,-1)$, depending upon the number contained in the current cell. We refer to the change in coordinates as a *movement*. So each movement is either $(1,1)$ or $(0,-1)$.

The total movement from the cell containing the number 1 to the cell containing the number $n+1$ is:

$$(n-1)(1,1) + (0,-1) = (n-1, n-2) \equiv (-1,-2) \pmod{n}.$$

Therefore, the total movement from the cell containing the number 1 to the cell containing the number $an+b+1$ is:

$$a(-1,-2) + b(1,1) = (-a+b, -2a+b) \pmod{n}.$$

Here a and b take values in the set S_n. It may be checked that as a and b take different values in the set, $an+b+1$ takes each value between 1 and n^2 once, with no repetitions or omissions. For each value of k between 1 and n^2, there corresponds a *unique* pair of values of a and b in the set S_n. (Indeed, $a = [k/n]$ and $b = k - 1 - an$.)

Moreover, the pairs $(-a+b, -2a+b) \pmod{n}$ are distinct; for if we have

$$(-a+b, -2a+b) \equiv (-a'+b', -2a'+b') \pmod{n},$$

then we readily get $a' \equiv a \pmod{n}$ and $b' \equiv b \pmod{n}$; and since $a, b, a', b' \in S_n$ it follows that $a' = a$ and $b' = b$. This proves the claim of distinctness. We infer that the algorithm described does indeed succeed in placing all the integers from 1 to n^2 in distinct cells.

The squares obtained for $n = 3$ and $n = 5$ are shown below.

8	1	6
3	5	7
4	9	2

17	24	1	8	15
23	5	7	14	16
4	6	13	20	22
10	12	19	21	3
11	18	25	2	9

Observe that in each square, the central number (respectively 5, 13) occupies the central cell. This turns out to be true always. Moreover, *numbers at equal distance from the central number are placed symmetrically about the centre*. To see what is meant by this, observe the placements of the pairs $(1, 25)$, $(2, 24)$, $(3, 23)$, and so on, in the fifth order square, above.

If we subtract the central number from each entry in the two squares shown above, we get the following:

3	−4	1
−2	0	2
−1	4	−3

4	11	−12	−5	2
10	−8	−6	1	3
−9	−7	0	7	9
−3	−1	6	8	−10
−2	5	12	−11	−4

The symmetry about the central square is very visible now. We shall show that the symmetry is an invariant feature.

- Since n is odd, the set $\{1, 2, 3, \ldots, n^2\}$ has a central number, namely $\frac{1}{2}(n^2 + 1)$; and as

$$\frac{n^2+1}{2} = \left(\frac{n-1}{2}\right)n + \frac{n-1}{2} + 1,$$

the a-, b-values referred to earlier, corresponding to $k = \frac{1}{2}(n^2 + 1)$, are:

$$a = \frac{n-1}{2}, \quad b = \frac{n-1}{2}.$$

Therefore, $-a+b=0$ and $-2a+b=-a\equiv(n+1)/2 \pmod{n}$. So the total movement from the location of the initial '1' is $(0,m+1)$. As the starting cell is $(m,2m)$, the location of the central number is:

$$(m,2m)+(0,m+1)=(m,3m+1)\equiv(m,m) \pmod{2m+1}.$$

So the central number comes in cell (m,m)—dead centre.

- Next, we show that the numbers that are symmetrical with respect to the central number $\frac{1}{2}(n^2+1)$ are placed symmetrically about the centre.

  Let k and k' be numbers that are symmetrical with respect to the central number $\frac{1}{2}(n^2+1)$, and let $k=an+b+1$ and $k'=a'n+b'+1$, where $a,b,a',b'\in S_n$; then it follows by symmetry that

  $$a+a'=n-1=2m, \quad b+b'=n-1=2m.$$

  (For example, for $n=5$, with central number 13, the a- and b-values corresponding to the pair 2, 24 are: $a=0$, $b=1$ and $a'=4$, $b'=3$.)

  The cells containing these numbers are $(-a+b+m,-2a+b+2m)$ and $(-a'+b'+m,-2a'+b'+2m)$ (both $\pmod{n}$). Observe that

  $$\begin{aligned}&(-a+b+m,-2a+b+2m)+(-a'+b'+m,-2a'+b'+2m)\\ =\;&(-a-a'+b+b'+2m,-2a-2a'+b+b'+4m)\\ =\;&(2m,2m),\end{aligned}$$

  implying that the two cells are symmetrically placed with respect to the central cell (m,m). This proves the claim made.

  From the central symmetry it follows immediately that both the diagonals have the correct sum, namely $n\times\frac{1}{2}(n^2+1)$.

- The verification that the square obtained by this method is a magic square is now easy. We only show a sample computation, for the middle column, for which $a=b$, implying that $2-a+b=-a\equiv n-a$. The pairs (a,b) corresponding to the middle column are

  $$(0,0),\quad(1,1),\quad(2,2),\quad\ldots,\quad(n-1,n-1),$$

  and the corresponding values of $an+b+1$ are

  $$1,\quad n+2,\quad 2n+3,\quad\ldots,\quad(an+a+1),\quad,\ldots,\quad n^2.$$

The sum of these numbers is:

$$\begin{aligned}\sum_{a=0}^{n-1}(an+a+1) &= n+(n+1)\sum_{a=0}^{n-1}a = n+\frac{(n+1)n(n-1)}{2}\\ &= \frac{n^3+n}{2} = \frac{1+2+3+\ldots+n^2}{n},\end{aligned}$$

and the sum is seen to be correct. So the algorithm works!

Further generalization is possible; see Chapter 7 of the book by W W Rouse Ball ([2]), or Chapter 2 of the book by I Niven and H S Zuckerman ([5]) for details.

7.7 Dudeney's problem

This section deals with a problem posed long ago by H E Dudeney: *Do magic squares exist containing only prime numbers?* His answer was 'yes', and he gave the following 3×3 square as an example:

67	1	43
13	37	61
31	73	7

Unfortunately, this square has a '1' in it, which is not generally regarded as prime. So, it is relevant to ask whether there exist such squares containing no '1'. Fortunately, we find the answer once again to be *yes*. The material that follows is from [4] (see the list of references in the next section).

The search technique The author uses an algorithm to find such squares. He denotes the (i, j) entry of the 3×3 square by $a_{i,j}$ and assigns the following expressions to the various $a_{i,j}$:

$a_{1,1} = (x+y)/2$	$a_{1,2} = 1$	$a_{1,3} = (2x-y+1)/2$
$a_{2,1} = x-y+1$	$a_{2,2} = (x+1)/2$	$a_{2,3} = y$
$a_{3,1} = (y+1)2$	$a_{3,2} = x$	$a_{3,3} = (x-y+2)/2$

Here, x and y are parameters and we let x and y assume prime number values with $x < y$. It is not hard to find the basis behind these assignments. If we list only the x-coefficients of the entries in the matrix, this is what we see:

1/2	0	1
1	1/2	0
0	1	1/2

Observe that this array has the magic-square property (its row-, column- and diagonal-sums are all 3/2). If we extract the y-coefficients of the entries in the same way, we get the following array:

1/2	0	−1/2
−1	0	1
1/2	0	−1/2

This array too has equal row-, column- and diagonal-sums (0 in each case). The same property holds for the constant term (the portion free of x and y), the magic sum being 3/2. It follows that the given array has the magic-square property for any x and y, the magic sum being $3(x+1)/2$.

A computer search is now made, trying out all permissible values for x and y (within reasonable limits, say $y < x+1 < 1000$), and checking whether the various $a_{i,j}$ are prime for these values. Unfortunately, no satisfactory magic square emerges from this attempt.

We now introduce another parameter, z, to which we give a fixed odd prime value, and we assign values to the $a_{i,j}$ as follows (the constant terms in the above array have all been magnified by z):

$a_{1,1} = (x+y)/2$	$a_{1,2} = z$	$a_{1,3} = (2x-y+z)/2$
$a_{2,1} = x-y+z$	$a_{2,2} = (x+z)/2$	$a_{2,3} = y$
$a_{3,1} = (y+z)/2$	$a_{3,2} = x$	$a_{3,3} = (x-y+2z)/2$

As earlier, this square has the magic-square property for all x, y and z, the magic sum being $3(x+z)/2$.

A computer search is made, as earlier, with x and y taking prime values ($z < y < z+x < 1000$). With $z = 3$, no satisfactory magic square emerges, but the assignments $z = 5$ and $z = 7$ each provide magic squares composed of prime numbers only. Here is a typical such square, with $z = 5$ and the magic sum equal to 177.

71	5	101
89	59	29
17	113	47

We have obtained an answer to Dudeney's problem.

EXERCISE

7.7.1 Write a computer program to perform the search referred to above, and obtain more such magic squares.

7.8 Curiosities

To conclude this chapter, here are three more problems (the first two should be regarded as mere curiosities and should perhaps not be taken too seriously).

EXERCISES

7.8.1 Does there exist a 3×3 magic square composed only of square numbers?

7.8.2 Construct a $n \times n$ anti-magic square based on the number set $\{1, 2, \ldots, n^2\}$.
(An 'anti-magic square' is one whose row-, column- and diagonal-sums are all *distinct* from one another. There exists no 2×2 anti-magic square based on the number set $\{1, 2, 3, 4\}$. Please justify this for yourself.)

7.8.3 *Multiplication of magic squares.* Let A and B be two magic squares of like order. Let them be regarded as matrices, and consider the array corresponding to the matrix $C = A \cdot B$. (Note that C is square.) Show that the row- and column-totals of C are all equal, but that C is not necessarily a magic square (its diagonal sums may be different).

EXAMPLE. Let A and B be given by:

$$A = \begin{pmatrix} 1 & 8 & 13 & 12 \\ 14 & 11 & 2 & 7 \\ 4 & 5 & 16 & 9 \\ 15 & 10 & 3 & 6 \end{pmatrix}, \quad B = \begin{pmatrix} 22 & 12 & 18 & 87 \\ 23 & 57 & 40 & 19 \\ 93 & 11 & 31 & 4 \\ 1 & 59 & 50 & 29 \end{pmatrix}.$$

Then the product $C = A \cdot B$ is given by:

$$C = \begin{pmatrix} 1427 & 1319 & 1341 & 639 \\ 754 & 1230 & 1104 & 1638 \\ 1700 & 1040 & 1218 & 768 \\ 845 & 1137 & 1063 & 1681 \end{pmatrix}.$$

Observe that the row- and column-sums of C are 4726. How is this number related to the magic sums of A and B? (Note that the diagonal-sums are *not* equal to this number.)

7.9 Postscript on the 3 × 3 Square

The following property of the 3×3 magic square that uses the digits 1, 2, ..., 8, 9 was discovered rather recently (see [6]); it certainly gives a new meaning to the 'square' in 'magic square'!

We read the digits *along the rows* (left to right as well as right to left), *along the columns* (top to bottom as well as bottom to top), and along the *broken diagonals* (SE–NW as well as SW–NE; here 'SE' means south-east, 'NW' means north-west, and so on). Each time we concatenate the digits together and read them aloud as base-10 numbers. The square is shown below.

6	1	8
7	5	3
2	9	4

Given below are the numbers we read.

- *Row numbers.* Left to right: 618, 753 and 294; right to left: 816, 357 and 492.
- *Column numbers.* Top to bottom: 672, 159 and 834; bottom to top: 276, 951 and 438.
- *Diagonal numbers*, SE–NW. South-east numbers: 654, 132 and 879; north-west numbers: 456, 231 and 978.
- *Diagonal numbers*, SW–NE. South-west numbers: 852, 174 and 639; north-east numbers: 258, 471 and 936.

So, in each of the four categories we have two sets of base-10 numbers, three in each set, the numbers in each set being the "reversals" of those in the other set. Now here comes a truly startling property:

$$\begin{aligned} 618^2 + 753^2 + 294^2 &= 816^2 + 357^2 + 492^2, \\ 672^2 + 159^2 + 834^2 &= 276^2 + 951^2 + 438^2, \\ 654^2 + 132^2 + 879^2 &= 456^2 + 231^2 + 978^2, \\ 852^2 + 174^2 + 639^2 &= 258^2 + 471^2 + 936^2. \end{aligned}$$

We can get two more such identities from the last two sets of numbers, by reading the numbers the other way around:

- *Diagonal numbers, SE–NW′*. South-east numbers: 654, 213 and 798; north-west numbers: 456, 312 and 897.
- *Diagonal numbers, SW–NE′*. South-west numbers: 852, 417 and 396; north-east numbers: 258, 714 and 693.

The identities we now get are:

$$\begin{aligned} 654^2 + 213^2 + 798^2 &= 456^2 + 312^2 + 897^2, \\ 852^2 + 417^2 + 396^2 &= 258^2 + 714^2 + 639^2. \end{aligned}$$

These identities certainly cry out for an explanation!

Generalization In fact, such identities hold for *any* 3×3 magic square. So, for the magic square

a	b	c
d	e	f
g	h	i

we have the identity

$$\overline{\mathsf{abc}}^2 + \overline{\mathsf{def}}^2 + \overline{\mathsf{ghi}}^2 = \overline{\mathsf{cba}}^2 + \overline{\mathsf{fed}}^2 + \overline{\mathsf{ihg}}^2,$$

and five other such identities. Here, of course, $\overline{\mathsf{abc}}$ represents the number $100a + 10b + c$, and similarly for the others.

We shall not prove the identity here; we invite the reader to do so. (Obviously, if we succeed in proving this identity, then the others will follow in a like manner.)

7.10 References

[1] Benjamin, A T and Yasuda, K. Magic "Squares" Indeed!, *American Mathematical Monthly*, Volume **106**, Number 2, February 1999, page 153.

[2] Ball, W W R, and Coxeter, H S M. *Mathematical Recreations and Essays*, University of Toronto, 1974.

[3] Gardner, Martin. *More Mathematical Puzzles and Diversions* Penguin, Chapter 12, pages 104–110.

[4] Fults, John Lee. *Magic Squares*.

[5] Niven, I, and Zuckerman, H. *An Introduction to the Theory of Numbers*, John Wiley, 1972.

[6] Pavlov, Tsvetan. A Problem By Dudeney, *Mathematics & Informatics Quarterly*.

[7] Srinivasan, P K. *Harvesting Magic Squares*, Allied Publishers Limited.

[8] Srinivasan, P K. *Mathematics and Magic Squares*, AMTI, Madras, India.

Chapter 8

Logic

A chapter devoted entirely to problems in logic may seem odd. Is not any mathematical problem ultimately a problem in logic? Does not the solution of any mathematical problem, however intricate, ultimately have to be presented logically,[1] completely? While this is undeniably so, there is nevertheless some virtue in identifying a class of problem known as *logic problems*. Here is the definition given by Martin Gardner in *More Mathematical Puzzles and Diversions*:

> A brain teaser that calls for deductive reasoning with little or no numerical calculation is ... labelled a logic problem.

The definition is clearly not be taken as absolute; the phrase "little or no calculation" is sufficiently vague to ensure an element of subjectivity in the classification. (Note: The term 'numerical' may be taken to include algebraic computations.)

Lewis Caroll (1832–1898) was a pioneer in the field of logic problems. He formulated a number of puzzles that require the solver to piece together a collection of disconnected statements and logically arrive at a single deduction. Caroll's examples all carry a touch of absurdity in them, and this of course adds greatly to the fun.

Another form, pioneered by Dudeney and popular to this day, is epitomized by the Smith-Jones-Robinson puzzle (named after

[1]Note that logic does not necessarily lead us inevitably to the solution of a problem; something more is needed, perhaps a flash of insight. However, the *presentation* of the solution must necessarily be logical and complete.

the characters appearing in the original problem). Boat-crossing puzzles are another popular form; perhaps the best known is the one featuring a wolf, a cabbage and a goat. In recent times, Raymond Smullyan has popularized a form of logic puzzle which involves self-reference (inspired, no doubt, by the work of the logician Kurt Gödel). We shall consider a few problems of each of these varieties.

8.1 Absurdities from Lewis Caroll

What conclusion follows logically from the statements given below?

(1) Babies are illogical.
(2) Nobody is despised who can manage a crocodile.
(3) Illogical persons are despised.

⋆ ⋆ ⋆

The problem is from Lewis Caroll, who clearly enjoyed devising such problems; many of them are to be found scattered through his all-time classics, *Alice in Wonderland* and *Through the Looking Glass*. In *Symbolic Logic*, Caroll presents an algebraic approach, using symbolic logic, of solving his puzzles.

Brief comments on symbolic logic The basic object that logic deals with is a *proposition*: a statement that asserts something. What is says is verifiably true or false (though one may not know which!).

Examples. : "The sun is shining"; "It rained last night"; "$3+5=8$"; $2\times 7=11$ (this is, of course, false). Note that questions are not propositions; nor are declarations of the type "I think that red is a nice colour".

Next, we have the logical connectives 'and', 'or' and 'implies'. These allow us to connect propositions together and arrive at *compound propositions*. The symbols used are '$\wedge$' for 'and' and '$\vee$' for 'or'; however, we shall avoid using these symbols here. Note that 'p and q' (written $p \wedge q$) is true iff *both* p and q are true, while 'p or q' (written $p \vee q$) is true iff at least one out of p, q is true (equivalently, $p \vee q$ is false iff both p and q are false). The 'implies' connective is denoted by '$\Rightarrow$'. The proposition '$p \Rightarrow q$'

is falsified if p is true whereas q is false (this is the only situation when it is falsified).

Some of the basic modes of reasoning are given below. *They form the 'substratum' of any proof in mathematics or in logic.*

(a) *Transitivity of the 'implies' relation.* If p, q and r are propositions, and if we have the two propositions

$$p \Rightarrow q, \qquad q \Rightarrow r,$$

then we can assert that $p \Rightarrow r$.

(b) *Law of the contrapositive.* If p and q are propositions, then the two propositions

$$p \Rightarrow q, \qquad \sim q \Rightarrow \sim p,$$

are logically equivalent to one another.

(c) *Law of contradiction.* If p and q are propositions, then from the two propositions

$$p \Rightarrow q; \qquad \sim q,$$

one can deduce $\sim p$.

Note that (b) and (c) say much the same thing. In addition to these three, we have the basic law known as the "law of the excluded middle"; namely, that $\sim(\sim p) = p$. Or, as Sherlock Holmes puts it in Conan Doyle's impeccable language,

> How many times have I told you, Watson, that when all else has been eliminated, what remains, however improbable, must be the truth ....

Solution of Caroll's problem Let us now apply these ideas to Caroll's problem about the illogical babies. We assign symbols to the various attributes as follows:

p : "is a baby";
q : "is illogical";
r : "is despised";
s : "can manage a crocodile".

Caroll's three statements may now formulated as follows.

$$p \Rightarrow q,$$
$$s \Rightarrow \sim r,$$
$$q \Rightarrow r.$$

The second statement may be stated in the contrapositive form as $r \Rightarrow \sim s$. So we get the following string of statements:

$$p \Rightarrow q, \quad q \Rightarrow r, \quad r \Rightarrow \sim s,$$

and we readily deduce that $p \Rightarrow \sim s$. In other words,

Babies cannot manage crocodiles!

★ ★ ★

A slightly more involved example is the following, again from Caroll.

(1) Nobody who really appreciates Beethoven fails to observe silence while the Moonlight Sonata is being played.
(2) Guinea-pigs are hopelessly ignorant of music.
(3) No one who is hopelessly ignorant of music ever observes silence while the Moonlight Sonata is being played.

We assign symbols to the various attributes as follows.

p : "is a guinea-pig";
q : "is hopelessly ignorant of music";
r : "observes silence while the Moonlight Sonata is being played"
s : "really appreciates Beethoven".

The statements now translate into the following:

$$s \Rightarrow r,$$
$$p \Rightarrow q,$$
$$q \Rightarrow \sim r.$$

Replacing the first statement by its contrapositive form, we get:

$$p \Rightarrow q, \quad q \Rightarrow \sim r, \quad \sim r \Rightarrow \sim s,$$

and we readily deduce that $p \Rightarrow \sim s$. In other words,

Guinea-pigs do not really appreciate Beethoven!

★ ★ ★

The reader may now wish to try the following, also from Caroll. (Remember that the given statements are all that may be used to draw valid inferences; in particular, no 'general knowledge' is to be used. Moreover, *all* the statements must be used.)

EXERCISES

8.1.1 Find a valid conclusion from the following propositions.

(1) No kitten that loves fish is unteachable.
(2) No kitten without a tail will play with a gorilla.
(3) Kittens with whiskers always love fish.
(4) No teachable kitten has green eyes.
(5) No kittens have tails unless they have whiskers.

8.1.2 Find a valid conclusion from the following propositions.

(1) Animals that do not kick are always unexcitable.
(2) Donkeys have no horns.
(3) A buffalo can always toss one over the gate.
(4) No animal that kicks is easy to swallow.
(5) No hornless animal can toss one over the gate.
(6) All animals are excitable, except buffaloes.

8.1.3 Find a valid conclusion from the following propositions.

(1) The only animals in this house are cats.
(2) Every animal, that loves to gaze at the moon is suitable as a pet.
(3) When I detest an animal, I avoid it.
(4) No animal is carnivorous, unless it prowls at night.
(5) No cat fails to kill mice.
(6) No animal ever takes to me, except those that are in this house.
(7) Kangaroos are not suitable as pets.
(8) None but carnivora kill mice.
(9) I detest animals that do not take to me.
(10) Animals that prowl at night always love to gaze at the moon.

8.2 'Smith-Jones-Robinson'

In this section we discuss a form of puzzle where, from a set of statements, some hidden information is sought. Dudeney's original problem, which has since spawned numerous descendants, concerned six persons aboard a train—two each with the names Smith, Jones and Robinson. From various clues provided the solver must deduce their identities. Here is a typical problem of this genre, based on a problem from Degrazia in *Math Tricks, Brain Twisters and Puzzles*.

Problem *On their way to a conference, six writers—a short-story writer, a historian, a humorist, a novelist, a playwright and a poet—meet in a train. They are seated opposite one another, three on each side. Each writer has a different speciality. Their names are A, B, C, D, E and F, but not necessarily in that order. Each writer has brought along one of his own books and given it to a colleague to read, so each one is deep in a book written by one of the others. The following facts are observed.*

1. *A is reading a collection of short stories. He is seated is a corner, and has no interest in history.*
2. *B is seated between the author of short stories and the humorist, and he is the brother-in-law of the novelist.*
3. *C is reading the book written by the person seated opposite him.*
4. *D is reading a play, and he is seated opposite the novelist.*
5. *The writer of short stories is seated opposite the historian.*
6. *E is seated next to the playwright, and he is reading a humorous book.*
7. *F detests reading poems.*

What is each writer's speciality?

Solution The problem is indeed quite challenging! The biggest difficulty is finding a way of 'holding' the information in some convenient display. We opt to display the information concerning authorship and who-is-reading-what-book in one array, and information about the seating arrangement in another array. A question mark means "do not know as yet", and 'Y' and 'N' indicate 'yes' and 'no', respectively. Thus, the 'N—Y' in the box next to A indicates that A is not the writer of short stories but is reading a book of short stories. In the second display, the notation ? || short story || ? indicates that the writer of short stories is seated in the 1st column, 1st row, but his identity is not known (as yet), nor what book he is reading.

	Short story	History	Humor	Novel	Play	Poetry
A	N—Y	N—N	?—N	?—N	?—N	?—N
B	N—N	?—?	N—N	N—?	?—N	?—?
C	?—N	?—?	?—N	?—?	?—N	?—?
D	?—N	?—N	?—N	N—N	N—Y	?—N
E	?—N	?—N	?—Y	?—N	N—N	?—N
F	?—N	?—?	?—N	?—?	?—N	?—N

The seating arrangement

? || short story || ?	B || ? || ?	? || humorist || ?
? || historian || ?	? || ? || ?	? || ? || ?

The displays shown above summarize the information given in the problem. Observe that B has been placed in the middle of the 1st row. Also, the writer of short stories, the historian and the humorist have been placed in definite slots. There is clearly no loss in generality in doing this.

We now argue as follows. The reader is urged to make copies of the above grids and to follow the thread of the argument on the grids with the help of a pencil (and eraser!). For convenience we refer to the columns as Col1, Col2 and Col3, and to the rows as Row1 and Row2. (Note: Once a 'Y' has appeared in any box, all the corresponding places in the boxes to its right or below it can be filled with 'N'.)

- A is not in Col1, as he is reading short stories but has no interest in history; ∴ he is in Col3 (remember that he is seated in a corner).

- D too is not in Col1, as he is seated opposite the novelist. If D were in Col2, then B would be the novelist, which is false as per the information given; ∴ D is in Col3. So both A and D are in Col3.

- Since the humorist is in Row1, and D is seated opposite the novelist, he must be in Row1; ∴ A is in Row2. So D is the humorist; A is the novelist.

- *Suppose* that C were in Col1; then he must be in Row1 (he cannot be in Row2 because he is reading the book written by the person seated opposite him, and the short story is being read by A). Next, E must be in Col1 (he cannot be in Row2/Col2 as he is next to the playwright); so the playwright must be in Row2/Col2, and he must be F (the only person left). This means that B is reading poetry (as F does not read poetry), and he must be the poet as well. This is absurd, as none of the writers is reading his own book.

It follows that C is *not* in Col1. Therefore, C is in Row2/Col2.

- *Suppose* that E were in Row1/Col1; then the playwright is in Row1/Col2, C is in Row2/Col2 and is reading a play. But we know that it is D who is reading the play—a contradiction.

- It follows that E cannot be in Row1/Col1. Since C too is not in this location, it must be F who is seated there (in Row1/Col1). This, in turn, means that E is in Row2/Col1. This means that C is the playwright and is reading poetry, and B is the poet.

So, the identities of all the persons are now revealed: A is the novelist, B is the poet, C is the playwright, D is the humorist, E is the historian and F is the writer of short stories.

We also have partial information on who-is-reading-what: A is reading a short story, C is reading poetry, D is reading a play, and E is reading a humorous piece. Note that we cannot ascertain what B and F are reading.

⋆ ⋆ ⋆

The fact that we are unable to ascertain what B and F are reading has a remarkable parallel in the work of Kurt Gödel in logic. This is the famous *undecidability theorem*; it states that in any axiom system, there are statements that can be made in the vocabulary of the system (relating to the elements of the system) whose truth or falsity cannot be decided by the axioms. *Any axiomatic system contains undecidable propositions.* In the problem just solved, the proposition "B is reading a novel" is undecidable, and so is its opposite: "B is reading a book on history". *The facts given in the problem do not tell us enough to be able to decide which of these two propositions is true and which one is false.*

We shall conclude this section with some problems.

Exercises

8.2.1 (This is from Erroll's book, *The Corgi Book of Problems.*) In a pub, five men (A, B, C, D and E) are sitting in a row along the counter. The following points are observed.

1. A man in a brown suit sits next to B.
2. A man on C's right drinks port.
3. A man who drinks whisky has red tie.
4. A wears brown suit and does not drink ale.
5. A man wearing Guard's tie drinks gin.

6. A gold-coloured tie is won by a man who drinks ale.
7. A man on B's right wears a blue suit.
8. D is not wearing a pin-stripe suit but is sporting a club tie.
9. A man in a grey suit has a black tie.
10. A man wearing a club tie is next to the man drinking ale.
11. A man drinking ale is next to the man wearing a blazer.
12. B drinks Guiness and does not have a red tie.
13. A man wearing a pin-stripe suit is drinking whisky.

Deduce what each man is wearing and what he is drinking!

8.2.2 [*] (Believe it or not, this problem was devised by the great British astrophysicist Arthur Eddington!)

Shown below is a cricket score card (batting scores first, then the bowling analysis).

A	B	C	D	E	F	G	H	I	J	K	Extras	Total
6	8	6	6	5	4	7	11	6	0	1	0	60

Bowler	Overs	Maidens	Runs	Wickets
P	12.1	2	14	8
Q	6	0	15	1
R	7	5	31	1

Further, the following has been recorded:

1. The score was made entirely from 4s and 1s.
2. There were no catches or short runs.
3. No-one was run out.
4. Q and R each had just one spell of bowling.
5. P bowled the first ball of the innings to A.
6. Q was the other opening bowler.

Of course, each over consists of six balls.

From this information, we are to deduce: (a) the identities of the batsman gotten out by Q and R (b) the identity of the 'Not Out' batsman (c) the score at the fall of each wicket.

8.2.3 [*] This problem has been adapted from one asked in the Australian 'Maths Challenge' of 1999. (It was devised by Andy Edwards, who is well-known for such devilish concoctions.)

In a hockey league tournament with five teams, each team plays each other team the same number of times. A win earns a team a certain number of points; so does a draw, but not as much as a win. Each of these numbers is a positive integer. A loss earns a team zero points. The end-of-league scores have been displayed below, with some information missing. The teams have been listed in decreasing order of total number of points scored (team A has the highest score and E has the lowest score).

Team	# Wins	# Losses	# Draws	Total Score
A	8			
B		5		
C	5			
D		6	1	43
E				22

It is known that no two teams got the same total score, and that except for E, every team got an odd total score.

Complete the score card, and find the number of points awarded for a win and for a draw!

8.3 Truth and lies

In recent times Raymond Smullyan, a magician-turned-professional mathematician, has popularized a form of puzzle involving certain island inhabitants who are invariably truthful or who invariably lie. Those who are always truthful are called the *knights*, and those who always lie are called the *knaves*. Here are two typical problems of this genre.

1. There are two persons, A and B, each a knight or a knave, and A says: "At least one of us is a knave". What are A and B?

2. There are three persons, A, B and C, each a knight or a knave, and A and B make the following statements.

A: "All of us are knaves."
B: "Exactly one of us is a knight."
What are A, B and C?

Solution to Problem 1 If A were a knave, then his statement would be false, implying that both A and B are knights; but this is absurd (A cannot be both a knight and a knave). Therefore, A is a knight, and so his statement is true, implying that B is a knave.

Solution to Problem 2 If A's statement were true, then A would have to be a knight—in contradiction with his own statement; therefore, A's statement is false, and A is a knave. Thus, there is at least one knight in the group.

If B's statement were false, then there have to be exactly two knights in the group, and these have to be B and C; again there is self-contradiction. Therefore, B's statement is true, and B is a knight. This means that C is a knave.

⋆ ⋆ ⋆

Smullyan also gives elaborate case-histories compiled by the energetic Inspector Craig of Scotland Yard. Here are two such cases.

1. A store has been raided and the loot driven away in a car. Three well known criminals – A, B and C – are brought to the Yard for questioning. The following facts become known.

   (a) No one other than A, B, C was involved in the robbery.
   (b) C never does a job without using A (and possibly others) as an accomplices.
   (c) B does not know how to drive.

   Is A innocent or guilty?

2. In another such robbery, three other criminals – P, Q and R – are brought for questioning. It is found that:

   (a) If P is guilty and Q is innocent, then R is guilty.
   (b) R never works alone.
   (c) P never works with R.

(d) No one other than P, Q, R is involved in the robbery, and at least one of them is guilty.

Which ones of P, Q, R are necessarily guilty?

Solution to Problem 1 If B were innocent, then at least one of A, C is guilty (someone must be guilty!); and if B is guilty, then at least one of A, C is guilty as B cannot drive. So in any case, at least one of A, C is guilty.

If C is innocent, then A must be guilty; and if C is guilty then too A is guilty, as C never works without A as an accomplice. So in any case, A is guilty.

COMMENT. The information does not reveal whether B and C are guilty.

Solution to Problem 2 *Suppose* that Q is innocent. Then if P is guilty, so is R, via (a); but this requires P and R to work together, which goes against (c); so P cannot be guilty. So *if* Q is innocent, then so is P, implying that R is guilty; but this goes against (b) which states that R never works alone. So there is self-contradiction here, and we infer that Q is guilty.

COMMENT. The information does not reveal whether P and R are guilty.

Exercises In the first two problems, we revisit the island inhabited by the knights and the knaves; but there is a difference. Since the time of the previous visit the *normals*, who are distinguished by the fact that they sometimes lie and sometimes tell the truth, have joined the knight and knaves on the island. They promise to add zest to the island life!

8.3.1 Three people, A, B and C, of whom one is a knight, one a knave and one a normal, though not necessarily in that order, make the following statements:

A: I am normal.
B: That is true.
C: I am not normal.

What are A, B and C?

8.3.2 In an unexpected move, the government passes a law that a knight can only marry a knave, and a knave can only marry a knave; so a normal can only marry a normal. The law has yet to be challenged.

Walking through the island we meet two couples, Mr/Mrs A, and Mr/Mrs B. They make the following statements:

Mr A says: "Mr B is a knight".

Mrs A says: "My husband is right; Mr B is a knight".

Mrs B says: "That's right. My husband is indeed a knight".

Identify each of the four persons. Which of the statements is true?

8.3.3 Three persons – A, B and C – are arguing with one another, each claiming to be the most intelligent. A friend, D, decides to test them. He shows them five stamps, three black and two white, and tells them that he plans to paste one stamp on each forehead, so that each person can see only the stamps on the other two foreheads. The first person to deduce the colour of his stamp is to be declared the most intelligent.

Unknown to them, D pastes a black stamp on each forehead. A few minutes later A announces: "I have a black stamp". How did he figure this out?

8.4 Boat crossings

The last genre of problems we shall consider deal with *boat crossings*. The ancestor of all these problems is the one about a man who reaches a river bank accompanied by a wolf, a goat and a cabbage. At the bank is a boat, and he must ferry himself and his companions across the river. Unfortunately, the boat can take only two entities at a time. Moreover, if he leaves the wolf with the goat on either bank, or in the boat, then the goat will get eaten up; and if he leaves the cabbage with the goat anywhere, it will similarly get eaten up. How does he get everyone across intact?

Write M, W, G, C to represent the man, wolf, goat and cabbage respectively; then the transfers may be done as follows.

M + G across; M back.

M + W across; M + G back.

M + C across; M back.
M + G across.

★ ★ ★

Here is another such puzzle. Three newly-married couples arrive at a river bank and find a boat which can hold at most two persons. They need to ferry themselves across. Unfortunately, each of the husbands is a terribly jealous person, and will not permit his wife to be in the company of any man—*unless he himself is present.* This rule holds throughout—on both banks, and in the boat itself; even 'transit contacts' are not permitted. How will the couples get across?

Denote the couples by (M1, W1), (M2, W2) and (M3, W3). The journeys may be accomplished as follows:

W1 + W2 across, W1 back.
W1 + W3 across, W1 back.
M2 + M3 across, M2 + W2 back.
M1 + M2 across, W3 back.
W1 + W2 across, W1 back, W1 + W3 across.

★ ★ ★

What happens if *four* couples find themselves in such a predicament? The answer is surprising: they simply cannot get across! The proof of impossibility is rather difficult, and the reader is invited to find it. However, the transfers become possible if slight modifications are made (see the exercise set below).

EXERCISES

8.4.1 Suppose that in the four-couples problem, the boat can hold three persons. The same conditions apply as earlier; the husbands are as jealous as the earlier lot! How do they get across?

8.4.2 In this version, the boat can hold only two persons, as earlier, but an island has conveniently appeared mid-stream, and the boat is permitted to deposit one passenger there, treating the island as a transit stop. How do the couples get across now?

8.4.3 [∗] Challenge! Prove that the original four-couples problem is not solvable.

8.5 References

[1] Caroll, Lewis. *Symbolic Logic* and *The Game of Logic* (published as one book by Penguin).

[2] Degrazia, Joseph. *Math Tricks, Brain Twisters and Puzzles*, Bell, 1981.

[3] Erroll, Jameson. *The Corgi Book of Problems*, Corgi, 1964.

[4] Gardner, Martin. *More Mathematical Puzzles and Diversions*, Penguin, 1961.

[5] Smullyan, Raymond. *What is the Name of this Book?*, Penguin, 1978.

Chapter 9

The Art of Counting

9.1 Introduction

Consider the following problem.

> **Diagonals of a polygon** Find a formula for d_n, the number of diagonals of a polygon with n sides. (*Example.* The number of diagonals is 2 when $n = 4$.)

It may be checked that $d_3 = 0$, $d_4 = 2$ and $d_5 = 5$; we seek a simple formula for d_n. How do we proceed? Two possible ways are given below. (Before reading further, the reader should try to tackle the problem independently.)

Solution 1 Let the vertices of the polygon be labelled in cyclic order: P_1, P_2, P_3, …, P_{n-1}, P_n (see Figure 9.1). Each diagonal of the polygon is of the form P_iP_j for some pair of indices $i < j$, except that if i, j differ by 1 or if $(i, j) = (1, n)$ then P_iP_j is a side of the polygon. The pairs (i, j) with $i = 1$ and $i < j$ are

$$(1,2),\ (1,3),\ (1,4),\ \ldots,\ (1,n-1),\ (1,n),$$

so there are $n - 1$ such pairs. The pairs (i, j) with $i = 2$ and $i < j$ are:

$$(2,3),\ (2,4),\ (2,5),\ \ldots,\ (2,n-1),\ (2,n),$$

so there are $n - 2$ such pairs.

Continuing, we find that there are $n - 3$ pairs $(3, j)$ with $3 < j$; $n - 4$ pairs $(4, j)$ with $4 < j$; and so on. The total number of pairs

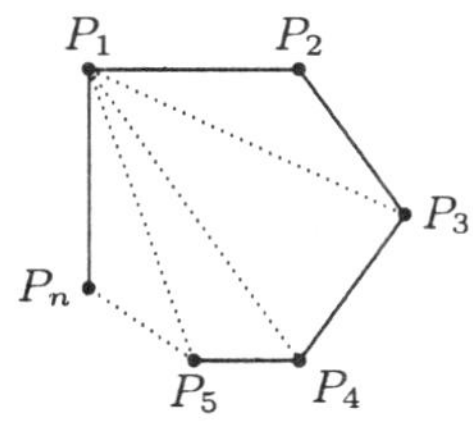

Figure 9.1. *Counting diagonals*

of all kinds is therefore S, where

$$S = (n-1) + (n-2) + (n-3) + \cdots + 3 + 2 + 1.$$

To find a formula for S is easy. The problem is a well-known one, and there is even a pretty story of Gauss's childhood days that goes with it. We simply reverse the order of the terms in the sum and write:

$$S = 1 + 2 + 3 + \cdots + (n-3) + (n-2) + (n-1).$$

Adding up the corresponding terms from the two expressions, we find that $1 + (n-1) = n,\ \ 2 + (n-2) = n,\ \ldots$, and so:

$$2S = \underbrace{n + n + n + \cdots + n + n + n}_{n-1 \text{ times}}.$$

Therefore, $2S = (n-1)n$ and $S = n(n-1)/2$. So, we have arrived at a formula for S. Now observe that d_n is $S - n$, for in the count for S we included the n sides (these correspond to the n pairs $(1,2)$, $(2,3)$, $(3,4)$, $\ldots$, $(n-1,n)$, $(1,n)$. It follows that:

$$d_n = \frac{n(n-1)}{2} - n = \frac{n(n-3)}{2}.$$

(*Check.* Consider a pentagon, with $n = 5$; the formula gives $d_5 = 5$, which is correct.) □

Solution 2 Here is another approach. The number of pairs of indices (i,j), where i and j belong to the set $\{1,2,3,\ldots,n-1,n\}$, is clearly n^2. (Each of i and j can take n different values, so (i,j) can take $n \times n = n^2$ different values.) We must discard all pairs of the form (i,i), as such pairs have no significance for us. There are n such pairs, so we replace the count by $n^2 - n$. In this count, pairs such as $(1,3)$ and $(3,1)$, or $(1,4)$ and $(4,1)$, are counted separately, though they refer to the same diagonal. As this is true for every pair, we must divide by 2; this changes the count to $(n^2 - n)/2$. Finally, we must subtract n from this expression, as the n sides of the polygon are included in the count (corresponding to pairs such as $(1,2)$, $(2,3)$, and so on). This yields:

$$d_n = \frac{n^2 - n}{2} - n = \frac{n(n-3)}{2},$$

as earlier. □

A party problem Now, consider the following problem.

> I go to a party where 5 other persons are assembled. Show that there are 3 amongst us, who all know one another *or* who are all unknown to one another.

(Observe that no 'counting' is being done in this problem.)

Solution to the party problem Write x for the number of persons I know at the party, and y for the number I do not know. Then $x + y = 5$, so the larger of x and y is 3 or more (if x, y were both less than 3, then $x + y$ would be less than 5). Suppose that $x \geq 3$. This means that I know at least 3 persons at the party, say A, B and C. Let M represent myself.

Now, consider the relationships between A, B and C. If any two of them know one another, say A and B, then amongst the three persons M, A, B, each pair knows one another, so we have at hand a threesome of the required sort. On the other hand, if no two of A, B, C know one another, then once again we have a threesome of the required sort (A, B, C themselves constitute the threesome). Similar reasoning works for the case when $y \geq 3$. (The reader should complete the proof.) □

Historical Comments 'Counting' forms the subject matter of this chapter, and it is a topic with curious origins. According to generally accepted folklore, the birth of the probability theory and combinatorics derives from the universal fascination that card- and dice-games hold for human beings. Historically, the science

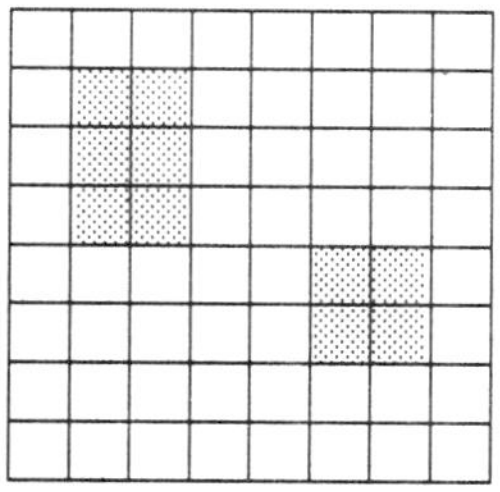

Figure 9.2. *Squares and rectangles on a chessboard*

of counting was born in a correspondence between Blaise Pascal, the philosopher–mathematician after whom 'Pascal's triangle' is named, and Pierre de Fermat, the jurist–mathematician after whom Fermat's Last Theorem is named. A shadowy figure who often features in the correspondence is the gambler Chevalier de Mère, who seems to have communicated many of his findings (empirically discovered, in the gambling dens of Paris!) to Pascal and Fermat.

However, we must note that Indian mathematicians of the 1500s and 1600s, notably Mahāvira and Bhāskarāchāryā, had found the basic results of combinatorics on their own. In Bhāskarā's famous book, *Lilāvati*, there is a clear statement of the formula for counting the number of permutations of a set of symbols. Here is a typical problem treated by him: *A number has 5 digits and the sum of the digits is 13. If 0 is not one of its digits, how many such numbers are possible?* Bhāskarā gives the answer as 495.

In modern times, the mathematician-educator George Polya and the 'travelling mathematician' Paul Erdős have elevated the art of counting to new heights. We shall offer some of their findings later in this chapter.

One of the best ways to get the flavor of any field of mathematics is by studying the problems posed in it, so we start by listing a few typical counting problems.

1. The usual 8×8 chessboard has 64 1×1 squares. It also has many squares measuring 2×2, 3×3, $\ldots$, formed by the lines of the chessboard. How many squares are there in all on the board, *of all possible sizes*? (One such square, of size 2×2, is shown shaded in Figure 9.2.)

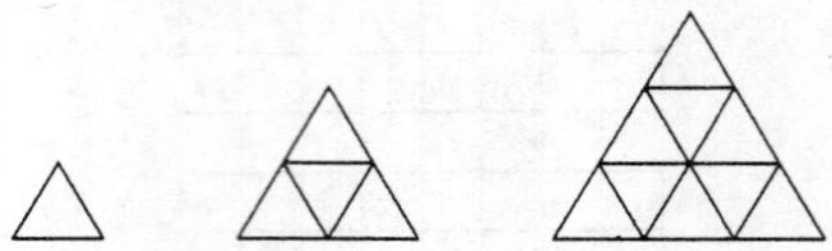

Figure 9.3. *Various triangular grids*

2. How many rectangles are there on a 8×8 chessboard, *of all possible sizes?* Here, the term 'rectangle' includes 'square'. (One such rectangle, of size 2×3, is shown shaded in Figure 9.2.)

3. Consider the triangular grids shown in Figure 9.3. The grid at the left (not much of a grid!—it has only one triangle) measures 1 unit on each side and of course it "contains" just 1 triangle.

   The grid in the middle measures 2 units on each side and has 5 triangles within it (4 small triangles, with a side of 1 unit, and 1 bigger triangle, with a side of 2 units), and the grid at the right, measuring 3 units on each side, has 13 triangles within it.

   How many triangles (of all possible sizes) are there in a triangular grid if the grid measures 10 units on each side?

4. Eight bridge players have assembled. They wish to break up into four teams of two each and then to pair up the teams to form two bridge-playing groups. In how many ways can the grouping be done?

5. Some girls have formed a human chain on a road. Amazingly, their names are all anagrams of the name DIVYA. Every possible anagram is represented by some girl in the chain, and no two girls have the same name. Moreover, the girls are lined up in the alphabetical order of their names, so the very first girl has the unusual name ADIVY, while the last one has the still more unusual name YVIDA. How many girls are there in the chain, and how many in-between DIVYA and VIDYA?

6. A positive integer is said to be *sorted* if its digits occur in

Figure 9.4. *How many shortest paths are there from A to B?*

strictly increasing or strictly decreasing order. Thus, the numbers 17, 239, 1458, 54, 932, ..., are sorted; whereas 254, 255, 655, 768, 213, ..., are not sorted. How many sorted integers are there in all? (The number is finite.)

7. **Taxi-Cab Geometry** In an earlier chapter, we studied taxi-cab geometry, where we are permitted to move only in the EW and NS (East-West, North-South) directions. I live at $A\,(0,0)$, and I need to visit $B\,(10,5)$, travelling along the shortest possible path. (So my journey is made up of steps of unit length, each pointing either due N or due E.) The length of a shortest path is 15 units. How many paths are there with this length? (See Figure 9.4.)

   Here is an example with smaller numbers: the shortest path from $A\,(0,0)$ to $C\,(2,2)$ has a length of 4 units, and one such shortest path (shown in Figure 9.4) is:

$$(0,0) \to (0,1) \to (1,1) \to (1,2) \to (2,2).$$

8. In how many ways can 8 rooks be placed upon a 8×8 chessboard, if no two rooks are to attack one another?

9. How many integers in the range 1 ... 2000 have no factor in common with 2000 (i.e., are *coprime* to 2000)?

10. I wish to paint each face of a cube with one of the colours: Red, White and Black. In how many distinguishable ways can I do this?

    (The key word here is *distinguishable*. Two colour schemes are said to be indistinguishable from one another if, given two cubes coloured according to the schemes, one can position them so that they look the same. This type of problem was studied in great detail by Polya.)

Did you solve any of the above problems? Here are the answers, for comparison: (1) 408 (2) 1296 (3) 315 (4) 315 (5) DIVYA is #34 and VIDYA is #68 in the line; there are 33 girls between them. (6) 1525 (7) 252 (8) 40320 (9) 800 (10) 57.

⋆ ⋆ ⋆

The reader encountering these problems for the first time may well find it hard to solve them. They belong to a branch of mathematics called *combinatorics*. Broadly speaking, combinatorics deals with arrangements, configurations, finite systems (more generally, discrete systems), counting, and so on. This definition perhaps does not say much, but to give an exact definition is difficult, and may not serve much purpose. *Counting*, or the art of enumeration, forms an important part of combinatorics. The problems listed above are of an enumerative nature, while the 'six persons at a party' problem solved earlier is not of this type. Here are some more problems of the non-enumerative type.

11. I colour each element of the set $S = \{1, 2, 3, \ldots, 8, 9\}$ either Red or Green. Show that no matter how I do this, I can select from S some three integers a, b, c that have a single colour and form an AP (this means that $a - b = b - c$). That is, we have a *3-term monochromatic arithmetic progression* somewhere in the set.
12. At a party of 250 mathematicians, each mathematician speaks one or more languages. It is found that for any two mathematicians, each speaks at least one language not spoken by the other. Show that there are at least ten different languages spoken at the party.
13. We are given a 9×9 array of white squares. Find the largest number n with the property that no matter which n squares of the array are coloured black, there remains an array of white squares measuring 1×4 or 4×1 (that is, there remains a white rectangle measuring 1×4).
14. There are n participants in a conference. Suppose that: (i) every two who know one another have no common acquaintance, and (ii) every two who do not know one another have precisely two acquaintances in common. Show that every participant in the conference is acquainted with the same number of persons.

15. At a meeting of 450 persons, each person slaps one other person. (This clearly refers to a political meeting!) Prove that one can select a group of 150 persons from amongst the 450, such that no member of the group has been slapped by any other member of the group.

16. At a party of 10 boys and 10 girls, each boy is acquainted with 2 girls and each girl is acquainted with 2 boys. Prove that one can pair up the boys and girls so that each person is paired with a person known to him or to her.

17. In a tournament of n participants played in the 'round-robin' style, each pair of participants plays one game that ends in a win for one of the players. (There are no draws.) Show that at the end of the tournament, either: (a) the league splits into two non-empty subsets A and B such that each player of group A has beaten each player of group B, or (b) the n players can be arranged in a cyclic sequence, in which each player has beaten the one behind him.

18. At a party with 17 persons, *each* pair of persons decide to play one of the following three games: backgammon, chess and billiards. Show that one can find three persons among them, say A, B, C, such that the pairs $\{A, B\}$, $\{B, C\}$ and $\{A, C\}$ all play the same game.

19. An $n \times n$ array of numbers, each number being drawn from the set $S = \{1, 2, 3, \ldots, 2n - 1\}$, is called a *silver array* if, for each $i = 1, 2, \ldots, n$, the ith row and i^{th} column together contain all the elements of S. Show that there do not exist any silver arrays for $n = 1997$.

    (Adapted from Problem 4 of the 38th IMO held at Mar del Plata, Argentina, in 1997.)

20. At a meeting of 9 mathematicians, each pair can converse in some language. No mathematician speaks more than three languages. Show that there is some language spoken by at least five mathematicians.

With this trial-by-fire introduction, we now proceed to make a leisurely excursion into the fascinating world of enumerative combinatorics.

9.2 The basic principles

We start by listing three basic principles of enumeration. These principles will seem utterly basic and elementary, but they have extremely wide applicability and can be used to establish a large number of propositions.

The notation we employ is standard: for any set A, the symbol $|A|$ denotes its cardinality (i.e., the number of elements in A). At times, we will also use the symbol $\#A$ to mean the same thing, at other times the notation $\text{card}(A)$. All the sets encountered here are finite. The empty set, which contains no elements, is denoted by $\emptyset$.

C1. (*The "Equality Principle"*) If the elements of two sets A and B can be put into 1–1 correspondence with one another, then $|A| = |B|$. (No one can disagree with this!)

C2. (*The "Addition Principle"*) For any two sets A and B, we have

$$|A \cup B| = |A| + |B| - |A \cap B|,$$

with similar statements for the cases when more than two sets are involved, e.g.,

$$|A \cup B \cup C| = \sum |A| - \sum |B \cap C| + |A \cap B \cap C|$$

(here, $\sum |A| = |A| + |B| + |C|$, $\sum |B \cap C| = |B \cap C| + |C \cap A| + |A \cap B|$).

EXAMPLE. Let $A = \{a, b, c, d, e\}$, $B = \{c, d, e, f, g\}$; then $A \cup B = \{a, b, c, d, e, f, g\}$, $|A| = 5$, $|B| = 5$, $|A \cup B| = 7$, $|A \cap B| = 3$. Observe that $7 = 5 + 5 - 3$.

C3. (*The "Multiplication Principle"*) If a set A can be put into a 1–1 correspondence with the Cartesian product $B \times C$ of two other sets B and C, then $|A| = |B| \times |C|$. (The term "Cartesian product" is defined below.)

• **Cartesian product** The *Cartesian product* $B \times C$ of two sets B and C, is the set $\{(b, c) \mid b \in B, c \in C\}$.

EXAMPLE. If $B = \{1, 2, 3\}$ and $C = \{k, l\}$, then

$$B \times C = \{(1, k), (2, k), (3, k), (1, l), (2, l), (3, l)\}.$$

Observe that $|B| = 3$, $|C| = 2$, $|B \times C| = 6 = 3 \times 2$.

From the multiplication principle, we deduce the following:

C3′. Suppose that $\mathcal{P}$ is a 2-step process such that the 1st step can be done in p different ways and the 2nd step in q different ways, *independent of the choice taken at the* 1st *step*. Then, $\mathcal{P}$ can be done in pq different ways.

Note that **C3′** has ready extensions to 3-step processes, 4-step processes, and so on.

9.3 Seven applications

As we have noted earlier, the principles listed in the preceding section have very wide applicability. Given in the subsections below are some results that can be proved using these principles.

9.3.1 Anagrams

Given a word having n distinct letters, its letters can be permuted (i.e., rearranged) in $n!$ distinct ways, where $n!$ (pronounced 'n factorial') is defined by

$$n! = n \times (n-1) \times (n-2) \times \cdots \times 2 \times 1.$$

So, $3! = 6$, $4! = 24$, $5! = 120$, $6! = 720$, and so on. By convention, $0!$ is given the value 1.

EXAMPLE. The letters of the 3-letter word MAN can be arranged in 6 ways:

MAN, MNA, AMN, ANM, NAM, NMA

(note that $6 = 3 \times 2 \times 1$), and the letters of MAIN can be arranged in 24 ways (the reader should list these 24 ways).

If repetition of letters occurs in the word, then this must be taken into account; the formula is different. Take for instance the word GOOD. We find that its letters can be arranged in 12 ways:

GOOD, GODO, GDOO, DGOO, DOGO, DOOG,
ODOG, ODGO, OODG, OGOD, OGDO, OOGD.

The general result is that if a word with n letters has a repeated letters of one kind, b repeated letters of another kind, c repeated letters of a third kind, and so on, then the number of ways in which its letters may be arranged is

$$\frac{n!}{a! \times b! \times c! \times \cdots}.$$

EXAMPLE. For the word GOOD the number of arrangements is $4!/2! = 12$, as seen above. For the word MATHEMATICAL the corresponding number is $12!/(2! \cdot 2! \cdot 3!) = 19958400$.

COMMENT. The factorial numbers increase at a terrific rate. Thus, 20! has 19 digits, 100! has 158 digits, and 1000! has 2568 digits (all in base-10).

9.3.2 Subsets–1

For a set A of cardinality n, the number of subsets of A with cardinality r is $\binom{n}{r}$ (pronounced 'n choose r'), where

$$\begin{aligned}\binom{n}{r} &= \frac{n!}{r!(n-r)!} \\ &= \frac{n}{1} \cdot \frac{n-1}{2} \cdot \frac{n-2}{3} \cdots \frac{n-r+1}{r}.\end{aligned}$$

EXAMPLE. If A is the 5-element set $\{a, b, c, d, e\}$, then the number of 2-element subsets of A is 10, the subsets being: $\{a, b\}$, $\{a, c\}$, $\{a, d\}$, $\{a, e\}$, $\{b, c\}$, $\{b, d\}$, $\{b, e\}$, $\{c, d\}$, $\{c, e\}$. Observe that, by definition,

$$\binom{5}{2} = \frac{5!}{2!\,3!} = \frac{120}{2 \times 6} = 10.$$

9.3.3 The binomial theorem

Let n and r be integers, with $0 \leq r \leq n$. Then the coefficient of x^r in the expansion of $(1+x)^n$ is $\binom{n}{r}$. That is, the expansion of $(1+x)^n$ is given by

$$(1+x)^n = 1 + \binom{n}{1}x + \binom{n}{2}x^2 + \cdots + \binom{n}{n-1}x^{n-1} + x^n.$$

(The dots represent in-between terms. In summation notation we write: $(1+x)^n = \sum_{r=0}^{n} \binom{n}{r}x^r$. Observe that $\binom{n}{0} = 1 = \binom{n}{n}$, for all positive integers n.)

EXAMPLE. Consider the case $n = 6$; by long multiplication, we obtain

$$(1+x)^6 = 1 + 6x + 15x^2 + 20x^3 + 15x^4 + 6x^5 + x^6.$$

Observe that the coefficients of x^3 and x^4 are 20 and 15, respectively, and that

$$\binom{6}{3} = \frac{6!}{3!\,3!} = 20, \quad \binom{6}{4} = \frac{6!}{4!\,2!} = 15.$$

9.3.4 Subsets–2

Let A be a set with n elements. Then the number of subsets of A, of all possible cardinalities, is 2^n. (The count includes the empty set $\emptyset$ and the set A itself.)

EXAMPLE. The set $A = \{c, d\}$ has 4 subsets, namely $\emptyset$, $\{c\}$, $\{d\}$ and $\{c, d\}$; note that $4 = 2^2$.

9.3.5 A basic recursion relation

The symbol $\binom{n}{r}$ obeys the following relation:

$$\binom{n}{r} = \binom{n-1}{r} + \binom{n-1}{r-1}.$$

This is known as a *recursion relation* as it relates the counting coefficients of the nth order with those of the $(n-1)^{\text{th}}$ order.

EXAMPLE. With $n = 7$ and $r = 3$, we have $\binom{n}{r} = \binom{7}{3} = 35$; $\binom{n-1}{r} = \binom{6}{3} = 20$; $\binom{n-1}{r-1} = \binom{6}{2} = 15$; and $35 = 20 + 15$.

In principle one can use this relation in conjunction with the 'boundary conditions',

$$\binom{1}{r} = \begin{cases} 1 & \text{for } r = 0 \text{ or } 1, \\ 0 & \text{for all other } r, \end{cases}$$

to compute the value of $\binom{n}{r}$ for any n and r. For instance, we compute $\binom{6}{3}$ thus:

$$\binom{6}{3} = \binom{5}{3} + \binom{5}{2} = \binom{4}{3} + 2\cdot\binom{4}{2} + \binom{4}{1}$$

$$= \binom{3}{3} + 3\cdot\binom{3}{2} + 3\cdot\binom{3}{1} + \binom{3}{0}$$
$$= \cdots = 20.$$

Note that for the two terms on the right side of the recursion equation to be meaningful, we must have $0 \le r \le n-1$ and $0 \le r-1 \le n-1$, or, combining the two, $1 \le r \le n-1$.

To avoid having to repetitively specify the allowable domains for n and r, we adopt the following convention:

If r is non-integral or negative or strictly greater than n, then $\binom{n}{r} = 0$; if $r = 0$, then $\binom{n}{r} = 1$; and in all other cases,

$$\binom{n}{r} = \frac{n}{1}\cdot\frac{n-1}{2}\cdot\frac{n-2}{3}\cdots\frac{n-r+1}{r}.$$

EXAMPLE. We have:

$$\binom{4}{-1} = 0, \quad \binom{3}{1/2} = 0, \quad \binom{5}{6} = 0;$$
$$\binom{4.5}{2} = (4.5 \times 3.5)/2 = 7.875.$$

Note that in this more general context, $\binom{n}{r}$ is an *algebraically* defined quantity, which happens to count something when n and r are integers and r lies between 0 and n. In this book, however, we shall not concern ourselves with the non-counting meanings of the symbol.

9.3.6 Pascal's triangle

Consider the well-known 'Pascal Triangle' shown below.

$$\begin{array}{ccccccccccc}
 & & & & & 1 & & & & & \\
 & & & & 1 & & 1 & & & & \\
 & & & 1 & & 2 & & 1 & & & \\
 & & 1 & & 3 & & 3 & & 1 & & \\
 & 1 & & 4 & & 6 & & 4 & & 1 & \\
1 & & 5 & & 10 & & 10 & & 5 & & 1
\end{array}$$

The defining condition here is that each row begins and ends with a 1, and every number not at the end of any row is the sum of the two numbers immediately above it. When the array is put into a square shape as follows:

	$r \rightarrow$	0	1	2	3	4	5	6
	0	1						
	1	1	1					
	2	1	2	1				
$n \rightarrow$	3	1	3	3	1			
	4	1	4	6	4	1		
	5	1	5	10	10	5	1	

with rows and columns indexed as shown, the number occurring in Row $\#n$ and Column $\#r$ is $\binom{n}{r}$. (The blanks are taken to mean '0'.)

9.3.7 The multinomial theorem

For any positive integer n, consider the expansion of

$$(x+y+z+\cdots)^n.$$

It is found that each term in the expansion is of the form $x^a \cdot y^b \cdot z^c \cdots$, where $a, b, c, \ldots$ are non-negative integers with sum n. (So, each term has a degree of n.)

EXAMPLE. The expansion of $(x+y+z)^2$ is

$$x^2 + y^2 + z^2 + 2xy + 2yz + 2zx.$$

Observe that each term has a degree of 2, and that all possible terms of degree 2 occur in the expansion.

In general, the coefficient of $x^a y^b z^c \ldots$ in the expansion of $(x+y+z+\cdots)^n$ is given by the expression

$$\frac{n!}{a!\,b!\,c!\,\ldots},$$

and all such possible terms occur. In other words,

$$(x+y+z+\cdots)^n = \sum_{a+b+c+\cdots=n} \frac{n!}{a!\,b!\,c!\,\ldots} x^a\, y^b\, z^c \ldots.$$

This is the so-called *multinomial theorem*. The symbol used for the coefficient $n!/(a!b!c!\ldots)$ is

$$\binom{n}{a, b, c, \ldots},$$

and it is pronounced as 'n choose $a, b, c, \ldots$'.

EXAMPLE. Consider the expansion of $(x+y+z)^3$. Here, each term has the form $x^a\, y^b\, z^c$, where a, b, c are non-negative integers with sum 3. The possibilities for (a, b, c) are $(3,0,0)$, $(2,1,0)$, $(1,1,1)$ and their various permutations. The relevant coefficients are

$$\binom{3}{3,0,0} = \frac{3!}{3!\,0!\,0!} = 1, \quad \binom{3}{2,1,0} = \frac{3!}{2!\,1!\,0!} = 3,$$

$$\binom{3}{1,1,1} = \frac{3!}{1!\,1!\,1!} = 6.$$

Therefore, the expansion of $(x+y+z)^3$ is

$$x^3 + y^3 + z^3 + 3x^2y + 3xy^2 + 3x^2z + 3xz^2 + 3y^2z + 3yz^2 + 6xyz.$$

Each of these propositions can be deduced from the principles listed earlier (**C1**, **C2**, **C3**). It is clear that these principles are not to be underestimated![1]

9.4 Harder examples

9.4.1 Boys and girls

In this section, we shall study the following problem.

> **Problem** *In how many ways can m boys and n girls be seated in a row of $m+n$ chairs, if no two boys are to be adjacent to one another?*

Clearly, we must have $n \geq m-1$, i.e., $m \leq n+1$, or there will be no arrangements of the required type.

Students often experience some difficulty with this problem (and others of its kind). However, there is a way of thinking about it that allows for swift progress. Imagine that the n girls are standing in a line, with the boys waiting to join them. The boys must now position themselves in-between the girls, however with the constraint that *no more than one boy occupies any given space.* This way, the condition that no two boys must be adjacent to one another is automatically satisfied.

Since there are n girls, the number of spaces available is $n+1$. To see why, consider the case $n = 3$. If 'G' denotes a girl, and

[1] Partly in reaction to this situation, some mathematicians complain that combinatorics is a branch of mathematics with too few basic theorems and too many problems!

'–' a space, then the situation may be depicted thus: –G–G–G–. Observe that there are 4 spaces that the boys can occupy. The general case is visualized in a similar way. This justifies why with n girls there should be no more than $n+1$ boys, or $m \leq n+1$.

The boys must occupy these spaces, each boy to a distinct space, and there are $n+1$ different spaces available. Since m spaces have to be chosen, the number of ways of doing this is

$$\binom{n+1}{m}.$$

Now, the girls can arrange themselves in $n!$ ways, and the boys can occupy a given set of m spaces in $m!$ different ways. So, the required number of ways is

$$\binom{n+1}{m} \times n! \times m! = \frac{(n+1)!\,n!}{(n+1-m)!}.$$

To round off the analysis we ask for the corresponding formula for the case when the girls too are not permitted to be adjacent to one another. This means, of course, that the boys and girls must alternated in the seating arrangement. This can be accomplished only if m and n differ by no more than 1, i.e., $|m-n| \leq 1$. There are three possibilities:

- $m = n+1$ (there are more boys than girls): in this case, the line starts and ends with a boy, and the number of possible arrangements is simply $m! \times n!$.
- $m = n$ (there are equal numbers of boys and girls): in this case, the left-most member may be either a boy or girl, and the number of possible arrangements is $2m! \times n!$.
- $n = m+1$ (there are more girls than boys): in this case, the line begins and ends with a girl, and as earlier the number of possible arrangements is $m! \times n!$.

So, the required number is either $m!\,n!$ or $2\,m!\,n!$, depending on whether m and n are unequal or equal.

9.4.2 Bridge hands

In the card-game of Bridge, each player is dealt 13 cards. (The deck consists of 4 suits, $\heartsuit$, $\spadesuit$, $\diamondsuit$ and $\clubsuit$, each with 13 cards.) The set of cards dealt out is referred to as a *hand* or a *deal*. The problem we study here is the following.

Problem In how many ways can a player be dealt a hand of 13 cards, if all four suits are to be represented in the deal?

We shall solve the problem in two different ways. The first way uses the Addition Principle as applied to the four sets.

Let the sets A, B, C, D be defined as follows: A is the set of all 13-card deals containing no $\heartsuit$; B is the set of all 13-card deals containing no $\spadesuit$; C is the set of all 13-card deals containing no $\diamondsuit$; and D is the set of all 13-card deals containing no $\clubsuit$. Then, clearly,

$$|A| = |B| = |C| = |D| = \binom{39}{13},$$

since in each case we must deal 13 cards from a set of 39 cards; remember that 13 cards have been kept out from the deal. Next, observe that

$$|A \cap B| = |A \cap C| = \cdots = |C \cap D| = \binom{26}{13},$$

for in the intersection $A \cap B$ we have 13-card deals with no $\heartsuit$ and no $\spadesuit$. Therefore, 13 cards must be dealt from a deck containing only 26 cards; and similarly for the other cases. Finally, note that the three-way intersections are given by

$$|A \cap B \cap C| = |A \cap B \cap D| = \cdots = 1.$$

(There is clearly just one way of dealing the cards, if three of the suits have been kept out.) Also, $|A \cap B \cap C \cap D| = 0$.

The Law of Addition as applied to 4 sets A, B, C, D yields the following: $|A \cup B \cup C \cup D|$ equals

$$(|A| + |B| + |C| + |D|) - (|A \cap B| + \cdots + |C \cap D|) +$$
$$(|A \cap B \cap C| + \cdots + |B \cap C \cap D|) - |A \cap B \cap C \cap D|.$$

When applied to our situation, the formula yields:

$$\begin{aligned} |A \cup B \cup C \cup D| &= 4|A| - 6|A \cap B| + 4|A \cap B \cap C| \\ &= 4 \times \binom{39}{13} - 6 \times \binom{26}{13} + 4. \end{aligned}$$

Finally, note that if a deal belongs to $A \cup B \cup C \cup D$, then at least one suit is missing from it. Since the total number of possible deals is $\binom{52}{13}$, the required number of deals, in which *all* suits are represented, is

$$\binom{52}{13} - 4 \times \binom{39}{13} + 6 \times \binom{26}{13} - 4$$
$$= 602586261420 \approx 6.026 \times 10^{11},$$

after substantial calculation. (Actually, DERIVE was used to do the calculation. Calculating $\binom{52}{13}$ manually would sorely stretch the author's patience!) So, the required number is a bit more than 600 billion.

Alternative solution Interestingly, there is another way of arriving at the answer. The underlying logic is very different, but the method is also more tedious.

Suppose we decide to have a, b, c and d cards each of $\heartsuit, \spadesuit, \diamondsuit$ and $\clubsuit$, respectively. Then the number of deals possible is, clearly,

$$\binom{13}{a} \times \binom{13}{b} \times \binom{13}{c} \times \binom{13}{d}.$$

Obviously we must have $a, b, c, d \in \mathbf{N}$, $1 \leq a, b, c, d \leq 13$ and $a + b + c + d = 13$. So the required number of ways is

$$\sum \binom{13}{a} \times \binom{13}{b} \times \binom{13}{c} \times \binom{13}{d},$$

where the summation is over all quadruples (a, b, c, d) of integers such that $1 \leq a, b, c, d \leq 13$ and $a + b + c + d = 13$.

We may now wish to manually list all such quadruples and do the relevant computation. But this is tedious, because the number of quadruples satisfying the given condition is 220; a little too large for comfort!

However, the summation can be broken down as follows. Since each of a, b, c, d is a positive integer, we must have $a \leq 10$, $a+b \leq 11$ and $a + b + c \leq 12$. Let a be fixed; then b can be no more than $11 - a$. Let b be fixed, with $1 \leq b \leq 11 - a$. Then c can be no more than $12 - a - b$. Let c be fixed so that $1 \leq c \leq 12 - a - b$; then d gets fixed: $d = 13 - a - b - c$. So the quantity we must compute is

$$\sum_{a=1}^{10} \sum_{b=1}^{11-a} \sum_{c=1}^{12-a-b} \binom{13}{a} \times \binom{13}{b} \times \binom{13}{c} \times \binom{13}{13 - a - b - c}.$$

Quadruple	Number
(10, 1, 1, 1)	4
(9, 2, 1, 1)	12
(8, 3, 1, 1)	12
(8, 2, 2, 1)	12
(7, 4, 1, 1)	12
(7, 3, 2, 1)	24
(7, 2, 2, 2)	4
(6, 5, 1, 1)	12
(6, 4, 2, 1)	24

Quadruple	Number
(6, 3, 3, 1)	12
(6, 3, 2, 2)	12
(5, 5, 2, 1)	12
(5, 4, 3, 1)	24
(5, 4, 2, 2)	12
(5, 3, 3, 2)	12
(4, 4, 4, 1)	4
(4, 4, 3, 2)	12
(4, 3, 3, 3)	4

Table 9.1. *Listing the quadruples with a sum of 13*

With the sum expressed in this manner, it may be computed using a computer algebra package, e.g., DERIVE. The answer comes out exactly the same as earlier (a little in excess of 600 billion).

⋆ ⋆ ⋆

The reader may wonder where the figure of 220 comes from (for the number of quadruples (a, b, c, d) with $1 \leq a, b, c, d \leq 13$ and $a+b+c+d = 13$). One way of getting the number is to list all quadruples (a, b, c, d) with $a \geq b \geq c \geq d$ and $a+b+c+d = 13$, then count the number of ways in each quadruple can be permuted, and then add the resulting numbers. Table 9.1 displays the result. ("Number" refers to the number of ways in which the given quadruple can be permuted.)

Example. For the quadruple $(10, 1, 1, 1)$ the number of permutations is $4!/3! = 4$, for $(9, 2, 1, 1)$ the number of permutations is $4!/2! = 12$, and so on.) The total number is seen from the table to be 220.

⋆ ⋆ ⋆

However, there is a shorter and decidedly nicer way of arriving at the figure of 220; it runs along the lines of the method we used in the "Boys and Girls" problem.

Let 13 coins be placed in a line as shown below, forming a chain; 'C' represents a coin, and '—' represent the space between two adjacent coins:

C–C–C–C–C–C –C–C–C–C–C–C–C.

There are 12 spaces in all. By choosing 3 of them we break the chain into 4 parts, each containing 1 or more C. These numbers give rise to a quadruple of integers (a, b, c, d) with the given properties (i.e., $1 \leq a, b, c, d \leq 13$ and $a + b + c + d = 13$). Conversely, each such quadruple gives rise to one such choice.

EXAMPLE. The quadruple $(2, 3, 5, 3)$ corresponds to choosing the 2nd, 5th and 10th spaces, and so on.

So the number of such quadruples is the same as the number of ways of choosing 3 spaces from 12 spaces; that is, it equals

$$\binom{12}{3} = \frac{12 \times 11 \times 10}{1 \times 2 \times 3} = 220,$$

as we had obtained earlier.

9.4.3 Ordered partitions

In this section, we investigate a problem concerning *ordered partitions*.

Let a positive integer n be written as a sum of one or more positive integers in all possible ways, with the order taken into account. These different sums are the 'ordered partitions' of n.

EXAMPLE. The ordered partitions corresponding to $n = 3$ are: 3, $2+1$, $1+2$, $1+1+1$ (four ways in all), and the ordered partitions corresponding to $n = 4$ are: 4, $3+1$, $1+3$, $2+2$, $2+1+1$, $1+2+1$, $1+1+2$, $1+1+1+1$ (eight ways in all).

We write $f(n)$ to denote the number of ways that n can be so expressed; so $f_3 = 4$ and $f_4 = 8$. Note that f_0 is not defined by this description. However, for convenience we agree to define f_0 as 1.

We now pose the problem of finding a simple and 'nice' formula for f_n. The answer is a most pleasing one.

Claim *We have $f_n = 2^{n-1}$ for all $n \in \mathbf{N}$.*

Observe that the formula 'works' for the cases $n = 3$ and 4, and also for $n = 1$ (there is just one way, and $1 = 2^0$) and $n = 2$ (there are two ways, 2 and $1 + 1$, and $2 = 2^1$).

There are many ways of proving the formula, and we shall opt for an approach that is sometimes described as *combinatorial*.

⋆⋆⋆

Consider the various ordered partitions corresponding to a given positive integer n. Take any partition, and let its right-most number be k If we delete this number, then the sum of the remaining numbers is $n-k$. Therefore, *the sum so obtained is an ordered partition for* $n-k$. There is more that can be said: *every ordered partition for* $n-k$ *is obtainable in this way*. So there is an exact match (a "1–1 correspondence", to use the technical term) between the ordered partitions of n whose right-most number is k and the ordered partitions of $n-k$: each ordered partition of $n-k$ yields an ordered partition of n whose right-most entry is equal to k, and this accounts for *all* the ordered partitions of n whose right-most entry is k. This statement holds even when $k=n$; for there is obviously only one partition of n with right-most number n, and $f_0 = 1$ (recall that we had agreed earlier to define f_0 as 1).

This argument clearly shows that

$$f_n = f_0 + f_1 + f_2 + \cdots + f_{n-2} + f_{n-1}.$$

This equation holds for every possible value of n, so it also holds when we replace n by $n-1$. Therefore,

$$f_{n-1} = f_0 + f_1 + f_2 + \cdots + f_{n-2}.$$

Comparing the two equations, we immediately see that

$$f_n = 2f_{n-1}.$$

So f_n gets doubled with each increase of 1 in n. Since $f_1 = 1 = 2^0$, it follows that $f_2 = 2\times 2^0 = 2^1$, and this implies that $f_3 = 2\times 2^1 = 2^2$, which in turn implies that $f_4 = 2\times 2^2 = 2^3$; and so on, to infinity. Clearly, this chain of reasoning implies that $f_n = 2^{n-1}$, as required. □

This is an example of a *combinatorial proof*. We have succeeded in counting the ordered partitions—without really counting!

⋆ ⋆ ⋆

There are, of course, other ways of obtaining the same result. Here is a particularly elegant combinatorial approach. Once again, it runs along the lines of the method used in the "Boys and Girls" problem.

Let n coins be laid out in a row. Between each adjacent pair of coins a space is formed, so there are $n-1$ spaces in all. We

Partition	Subset
4	$\{\}$
$3+1$	$\{3\}$
$2+2$	$\{2\}$
$1+3$	$\{1\}$
$2+1+1$	$\{2,3\}$
$1+2+1$	$\{1,3\}$
$1+1+2$	$\{1,2\}$
$1+1+1+1$	$\{1,2,3\}$

Table 9.2. *Correspondence between the partitions and the subsets*

number them as shown below (coins are represented by circles).

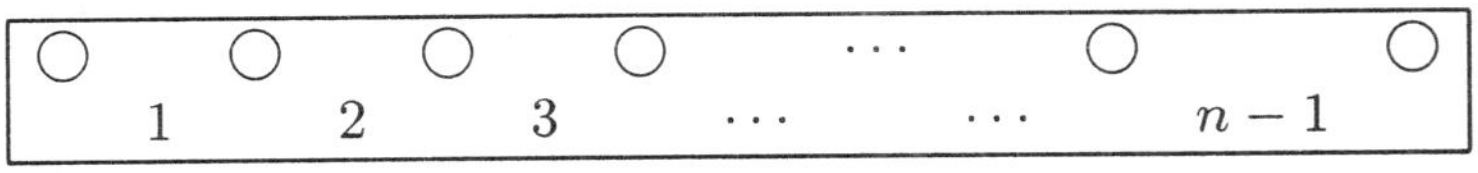

We claim that there is a 1–1 correspondence between the ordered partitions of n and subsets of the set $\{1,2,\ldots,n-1\}$. For, any subset of $\{1,2,\ldots,n-1\}$ partitions the row of coins in a natural manner, and this itself provides the desired ordered partition of n. To see explicitly how this works, consider the case $n=4$.

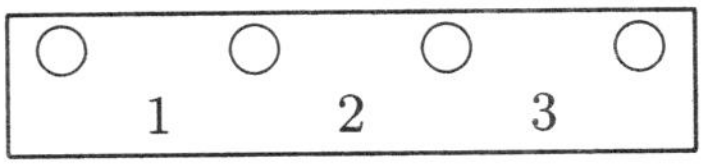

The correspondence is as shown in Table 9.2. Observe the inevitability and naturalness of the scheme; each subset creates its own ordered partition.

In the general case, the empty set corresponds to the partition n, and the full set corresponds to the partition $1+1+1+\cdots+1$ (consisting only of ones).

Therefore, the number of ordered partitions of n must be the same as the number of subsets of $\{1,2,3,\ldots,n-1\}$, and this number is clearly 2^{n-1}. We reach the same conclusion as we did earlier. Once again, we have succeeded in counting without doing any real hard work.

9.4.4 Derangements

Problem How many permutations are there of the following string on n symbols,

$$(1,2,3,\ldots,n-1,n),$$

in which every element occupies a place that is not its own?

A permutation in which each element gets shifted, with no element retaining its place, is known as a *derangement*.

EXAMPLE. Take $n = 4$; then $(2, 4, 1, 3)$ and $(4, 3, 2, 1)$ are derangements. On the other hand, $(2, 4, 3, 1)$ is not a derangement (as 3 is in the 3rd place).

Let D_n be the number of derangements on n symbols. Our task is to find a formula for D_n.

Obviously, $D_1 = 0$ and $D_2 = 1$ (the permutations of $(1, 2)$ are $(1, 2)$ and $(2, 1)$, and only the latter is a derangement). Next, $D_3 = 2$, for the permutations of $(1, 2, 3)$ are

$$(1, 2, 3),\ (1, 3, 2),\ (2, 3, 1),\ (2, 1, 3),\ (3, 1, 2),\ (3, 2, 1);$$

and of these only $(2, 3, 1)$ and $(3, 1, 2)$ are derangements.

Moving to the case $n = 4$, there are $4! = 24$ permutations of $(1, 2, 3, 4)$, which we may list as follows (without parentheses, for convenience):

$$\begin{array}{llllll} 1234, & 1243, & 1324, & 1342, & 1423, & 1432, \\ 2134, & 2143, & 2314, & 2341, & 2413, & 2431, \\ 3124, & 3142, & 3214, & 3241, & 3412, & 3421, \\ 4123, & 4132, & 4213, & 4231, & 4312, & 4321. \end{array}$$

It will be observed that there are 9 derangements, namely:

$$2143,\ 2431,\ 2413,\ 3142,\ 3412,\ 3412,\ 4123,\ 4312,\ 4321;$$

so $D_4 = 9$. Summarizing, we have $D_1 = 0$, $D_2 = 1$, $D_3 = 2$ and $D_4 = 9$. What is the pattern in the sequence?

Our strategy will be to find a formula connecting D_n with D_{n-1}, D_{n-2}, ... We reason as follows.

Take any derangement of $(1, 2, \ldots, n-1, n)$, and consider the element that occupies the 1st place; call it i. Since i belongs to the set $\{2, 3, \ldots, n-1, n\}$, there are $n-1$ possibilities for i. Now consider the position to which the '1' has been shifted; call it j. Obviously, $j > 1$.

EXAMPLE. Take $n = 5$. In the derangement $(3, 4, 1, 5, 2)$ we have $i = 3$, $j = 3$, and in $(5, 3, 4, 1, 2)$ we have $i = 5$, $j = 4$.

Two possibilities arise. (a) It may happen that $i = j$. This would mean that 1 and i have swapped places (as happens for $(3, 4, 1, 5, 2)$: 1 and 3 have swapped places). (b) Or it may happen that $i \neq j$. We shall consider both possibilities.

If $i = j$, that is, if 1 and i have swapped places, then the remaining $n-2$ elements, when considered as a whole, occupy the places that rightfully belong to them, but they must occupy them "all wrong"; i.e., the occupation must be a derangement. Since there are $n-2$ elements, this can be done is D_{n-2} ways. So for each $i = 2, 3, \ldots, n$, there are D_{n-2} derangements in which 1 and i swap places.

EXAMPLE. Take $n = 5$. Suppose that 1 and 3 have swapped places $(i = j = 3)$. Then the elements 2, 4 and 5, considered as a whole, must occupy the 2nd, 4th and 5th places, but each must go to a wrong place. Since D_3 is 2, there are 2 such derangements. They are the following: $(3,4,1,5,2)$ and $(3,5,1,2,4)$.

What happens if $i \neq j$? We know that i has occupied the 1st place, and now 1 has been denied the ith place; *it can move to any place except the* i^{th}. How many derangements are there satisfying this condition? We claim that the number is D_{n-1}.

EXAMPLE. Consider the case $n = 5$. Suppose that $i = 3$ (i.e., 3 occupies the 1st place). If 1 may not move to the 3rd place, it must move to the 2nd, 4th or 5th positions. The possibilities may be listed systematically as follows:

$$\begin{array}{lll} (3,1,2,5,4), & (3,1,4,5,2), & (3,1,5,2,4), \\ (3,4,5,1,2), & (3,5,2,1,4), & (3,5,4,1,2), \\ (3,4,2,5,1), & (3,4,5,2,1), & (3,5,4,2,1). \end{array}$$

Observe that there are 9 such derangements, the same number as D_4. This is not a coincidence; for with i occupying the 1st position, and 1 *not* occupying the ith position, the elements

$$1,\ 2,\ 3,\ \ldots,\ i-1,\ i+1,\ \ldots,\ n-1,\ n$$

must occupy positions normally reserved for the elements

$$i,\ 2,\ 3,\ \ldots,\ i-1,\ i+1,\ \ldots,\ n-1,\ n,$$

with the condition that 1 must not occupy the ith place, 2 must not occupy the 2nd place, 3 must not occupy the 3rd place; and so on. *This is a derangement on* $n-1$ *symbols*, only there has been some renaming of elements.

So with i in the 1st position and 1 not in the ith position, there are in all D_{n-1} derangements. This holds for each $i \in \{2, 3, \ldots, n-1, n\}$.

Summarizing, for each i in the set $\{2, 3, \ldots, n-1, n\}$ there are D_{n-2} derangements in which 1 and i swap places, and there are D_{n-1} derangements in which i comes to the 1st position and 1 moves to some position other than the ith. So in all there are $D_{n-2} + D_{n-1}$ derangements in which i moves to the 1st position. Since i can take $n-1$ possible values, we arrive at the equation

$$D_n = (n-1)\left(D_{n-2} + D_{n-1}\right).$$

This relation allows us to compute the 'derangement numbers' with great speed. Thus, starting with $D_1 = 0$ and $D_2 = 1$, we get $D_3 = 2 \times (0+1) = 2$, $D_4 = 3 \times (1+2) = 9$, $D_5 = 4 \times (2+9) = 44$, $D_6 = 5 \times (9+44) = 265$, and so on. Observe that the numbers increase very rapidly.

Even better, we can use the relation to obtain a 'closed-form' expression for D_n; we only need to exercise a bit of algebraic ingenuity. We first write the above relation, $D_n = (n-1)\left(D_{n-2} + D_{n-1}\right)$, as

$$D_n - nD_{n-1} = -\left(D_{n-1} - (n-1)D_{n-2}\right).$$

Observing the symmetry of form, we immediately infer that

$$\begin{aligned} D_{n-1} - (n-1)D_{n-2} &= -\left(D_{n-2} - (n-2)D_{n-3}\right), \\ D_{n-2} - (n-2)D_{n-3} &= -\left(D_{n-3} - (n-3)D_{n-4}\right), \end{aligned}$$

and so on.

Write, for convenience, a_n for $D_n - nD_{n-1}$; then we have

$$a_n = -a_{n-1} = a_{n-2} = -a_{n-3} = \cdots.$$

Since $a_2 = D_2 - 2D_1 = 1$, we deduce that $a_n = \pm 1$ for all n; and since a_2 is 1, we obtain $a_n = (-1)^n$. That is,

$$D_n - (n-1)D_{n-1} = (-1)^n.$$

This may be written in another way which is more useful. Transposing terms, we have, $D_n = nD_{n-1} + (-1)^n$. Dividing by $n!$, we get

$$\frac{D_n}{n!} = \frac{D_{n-1}}{(n-1)!} + \frac{(-1)^n}{n!}.$$

So, considering the quantities $D_n/n!$, we see that

$$\frac{D_1}{1!} = 0, \quad \frac{D_2}{2!} - \frac{D_1}{1!} = \frac{1}{2!}, \quad \frac{D_3}{3!} - \frac{D_2}{2!} = -\frac{1}{3!}, \quad \ldots,$$

and we quickly infer from this, by summation, that

$$\frac{D_n}{n!} = \frac{1}{2!} - \frac{1}{3!} + \frac{1}{4!} - \cdots + \frac{(-1)^n}{n!}.$$

It follows that

$$D_n = n!\left(\frac{1}{2!} - \frac{1}{3!} + \frac{1}{4!} - \cdots + \frac{(-1)^n}{n!}\right),$$

and thus we have our desired formula.

This formula yields a rather surprising conclusion. Consider the quantity

$$\frac{1}{2!} - \frac{1}{3!} + \frac{1}{4!} - \cdots + \frac{(-1)^n}{n!},$$

and call it c_n for convenience (note that c_n is defined only for $n \geq 2$; we may complete the definition by taking $c_1 = 0$). Then $c_2 = 1/2$, $c_3 = 1/3$, $c_4 = 3/8$, and so on. Continuing the computations and expressing the numbers as decimal fractions, we obtain the following:

n	3	4	5	6	7	8
c_n	0.333	0.375	0.367	0.368	0.368	0.368

It will be seen that for $n \geq 6$, we have $c_n \approx 0.368$; the value hardly changes at all as n increases! So we arrive at the following curious result:

For $n \geq 6$, approximately 36.8% of the $n!$ permutations of $(1, 2, 3, \ldots, n)$ are derangements, and this is so no matter how large n is.

Classification of permutations by fixed points To conclude, the following relation may be of interest:

$$n! = \binom{n}{0}D_n + \binom{n}{1}D_{n-1} + \cdots + \binom{n}{n-1}D_1 + \binom{n}{n}D_0,$$

where, for convenience, we define $D_0 = 1$.

EXAMPLE. Take $n = 4$. We know that $D_0 = 1$, $D_1 = 0$, $D_2 = 1$, $D_3 = 2$, $D_4 = 9$. Observe that

$$4! = (1 \times 9) + (4 \times 2) + (6 \times 1) + (4 \times 0) + (1 \times 1),$$

in agreement with the equation. The reader should verify the equation for $n = 5$.

To prove the relation, we classify the $n!$ permutations in terms of how many fixed points they have (a fixed point is any element that occupies its rightful place. Thus, taking $n = 5$, in the permutation $(2, 3, 1, 4, 5)$ there are two fixed points, namely 4 and 5).

Suppose that the number of fixed points is i. Since there are n elements to choose from, i fixed points may be chosen in $\binom{n}{i}$ ways. The remaining $n - i$ elements must form a derangement, as the permutation has exactly i fixed points. This means that they can be arranged in D_{n-i} ways. So the number of permutations with exactly i fixed points is

$$\binom{n}{i} D_{n-i}.$$

Since i can take values from 0 to n, and the total number of permutations is $n!$, we obtain

$$n! = \sum_{i=0}^{n} \binom{n}{i} D_{n-i},$$

as asserted.

9.5 Binomial identities

In this section, we offer a brief excursion into the rich subject of binomial identities. Our aim will be to provide, wherever possible, *non-computational proofs* of the identities (elsewhere, we shall refer to such proofs as *combinatorial*). What this means is that the two sides of the to-be-proved identity have to be given a combinatorial or 'counting' significance, following which equality is established by exhibiting a 1–1 correspondence between the configurations being counted. We offer a few examples below for illustration.

EXAMPLE 1. We start with the identity

$$\binom{n}{0}^2 + \binom{n}{1}^2 + \binom{n}{2}^2 + \cdots + \binom{n}{n}^2 = \binom{2n}{n},$$

or, more compactly, $\sum_{i=0}^{n} \binom{n}{i}^2 = \binom{2n}{n}$.

A sample numerical verification: for $n = 4$ the quantity on the left is $1^2 + 4^2 + 6^2 + 4^2 + 1^2 = 1 + 16 + 36 + 16 + 1 = 70$, and the quantity on the right is $\binom{8}{4} = 70$.

Here is the proof. From the set of n boys and n girls we must select a group of n individuals, with the boy-to-girl ratio not fixed in advance. Clearly, the number of possible groups is $\binom{2n}{n}$.

If, however, we decide to choose i boys and $n-i$ girls, for some fixed i, then the number of ways of making the selection is

$$\binom{n}{i} \times \binom{n}{n-i} = \binom{n}{i} \times \binom{n}{i} = \binom{n}{i}^2.$$

It follows readily that $\displaystyle\sum_{i=0}^{n} \binom{n}{i}^2 = \binom{2n}{n}$.

EXAMPLE 2. We shall show that

$$1 \cdot \binom{n}{1} + 2 \cdot \binom{n}{2} + \cdots + n \cdot \binom{n}{n} = n \cdot 2^{n-1},$$

or, more compactly, $\sum_{i=1}^{n} i \cdot \binom{n}{i} = n \cdot 2^{n-1}$.

Here is a non-computational proof. From a set of n persons, a group is to be formed with at least one member, with one person in the group designated as its leader. In how many ways can this be done? There are two ways of proceeding:

- Fix the group size i; choose i persons from the set, in $\binom{n}{i}$ ways; and then designate one of the chosen ones as the leader of the group. For fixed i this can be done in $i \cdot \binom{n}{i}$ ways. Therefore, the number of ways of doing the required task is $\sum_i i \cdot \binom{n}{i}$.

- Choose the leader in advance (this can be done in n ways), and then choose the remaining members of the group (this can be done in 2^{n-1} ways, as there are $n - 1$ persons to choose from, and *any* subset will do, even the empty set). Therefore, the number of ways of doing the required task is $n \cdot 2^{n-1}$.

It follows that $\displaystyle\sum_{i=1}^{n} i \cdot \binom{n}{i} = n \cdot 2^{n-1}$.

★ ★ ★

Small modifications yield other such identities with remarkable ease.

EXAMPLE. Suppose that it is specified that the group size be at least two, and that the leader have a designated deputy; then, by duplicating the argument given above, we arrive at the identity

$$\sum_{i=2}^{n} i \cdot (i-1) \cdot \binom{n}{i} = n \cdot (n-1) \cdot 2^{n-2}.$$

More generally, we have, for $k = 1, 2, 3, \ldots$:

$$\sum_{i=k}^{n} i \cdot (i-1) \cdot \cdots \cdot (i-k+1) \cdot \binom{n}{i} = n \cdot (n-1) \cdot \cdots \cdot (n-k+1) \cdot 2^{n-k}.$$

A similar argument establishes that for integers i, j, n with $0 \leq j \leq i \leq n$,

$$\binom{n}{i} \cdot \binom{i}{j} = \binom{n}{j} \cdot \binom{n-j}{i-j}.$$

EXAMPLE 3. We move to harder examples. Our next identity is inspired by Marta Sved's article *Counting and Recounting*:

$$\sum_{k=1}^{n} \sum_{i=0}^{k-1} \binom{k-1}{i} \binom{n-k}{i} = 2^{n-1}.$$

We consider subsets of the set $\{1, 2, 3, \ldots, n-1, n\}$, which have an *odd* number of elements. It is easy to show (and this is Problem 9.6.15) that the number of subsets with an odd number of elements is the same as the number of subsets with an even number of elements; both cardinalities are equal to 2^{n-1}. Any such subset has a uniquely defined 'middle' number (the *median* of the subset). Let the median be k and let the subset have $2i+1$ elements, i elements on either side of k (obviously we must have $1 \leq k \leq n$ and $0 \leq i \leq k-1$). The number of choices for the elements less than k is $\binom{k-1}{i}$, and the number of choices for the elements greater than k is $\binom{n-k}{i}$, so the number of odd-cardinality subsets with median k is

$$\binom{k-1}{i} \times \binom{n-k}{i}.$$

It follows, by summation over k, that

$$\sum_{k=1}^{n} \sum_{i=0}^{k-1} \binom{k-1}{i} \binom{n-k}{i} = 2^{n-1}.$$

EXAMPLE 4. Here is a particularly beautiful and unusual identity, which, too, is considered by Sved:

$$n^n = \sum_{i=1}^{n} \binom{n}{i} \times i! \times i \times n^{n-i-1}.$$

A sample verification, with $n = 4$: the quantity on the left is $4^4 = 256$, and the quantity on the right is $64 + 96 + 72 + 24 = 256$.

To prove the identity, we consider n coins, numbered 1, 2, 3, ..., n, and n boxes, similarly numbered. The coins have to be placed in the boxes, with no restrictions. Clearly, the number of ways in which this can be done is n^n, as each coin can go into any box.

On the other hand, for each possible placement, a number i can be *uniquely* identified such that coins 1, 2, ..., i go into different boxes, whereas coin $(i+1)$ goes into a box already occupied ($i = n$ means that each coin is in a different box; $i = 1$ means that coin 2 is in the same box as coin 1). For a fixed i, how many possible selections are there? The i boxes occupied by coins 1, 2, ..., i can be chosen in $\binom{n}{i}$ ways and occupied in $i!$ ways, implying a total of $i! \times \binom{n}{i}$ ways. Coin $(i+1)$ now has i boxes to choose from. Finally, coins $(i+2)$, $(i+3)$, ..., n have unrestricted access to the boxes, and can be placed in n^{n-i-1} ways. Therefore, the number of possible placements is

$$i! \times \binom{n}{i} \times i \times n^{n-i-1}.$$

It follows that

$$n^n = \sum_{i=1}^{n} \binom{n}{i} \times i! \times i \times n^{n-i-1}.$$

Since

$$\binom{n}{i} \times i = \binom{n-1}{i-1} \times n,$$

the identity may also be written as

$$n^n = \sum_{i=1}^{n} \binom{n-1}{i-1} n^{n-i}\, i!,$$

in which form it is known as Riordan's identity.

* * *

The reader is invited to tackle the following identities, with the cautionary warning that they are rather difficult.

EXERCISES

9.5.1 Show that

$$\sum_{i=1}^{n} i \binom{n}{i}^2 = n \binom{2n-1}{n-1}.$$

9.5.2 Show that

$$\sum_{i=0}^{n} 2^{n-i} \binom{n+i}{i} = 2^{2n}.$$

9.5.3 Show that

$$\sum_{i=0}^{n} \binom{2i}{i} \binom{2n-2i}{n-i} = 2^{2n}.$$

9.5.4 Show that if $0 < m < n$, then

$$\sum_{k=m}^{n} (-1)^k \binom{n}{k} \binom{k}{m} = 0.$$

9.5.5 Prove Abel's identity:

$$\sum_{i=1}^{n-1} \binom{n}{i} i^{i-1} (n-i)^{n-i-1} = 2(n-1)\, n^{n-2}.$$

★ ★ ★

We shall encounter the binomial coefficients again briefly in the next chapter (*Number Theory*), where we study their arithmetical properties.

9.6 Problems

In the problems listed below, we do not use the approach used elsewhere in this book—that of quoting a result and then asking for its proof. Instead, we opt for an investigative approach: that of experimentation with smaller configurations (smaller numbers, fewer variables, . . . ; i.e., with size and complexity kept to a modest level), analysis of the patterns involved, and so on. This approach may, at times, be the only one available when the problem is very complex. (A suggestion of Polya's is worth quoting here: if a problem seems too hard, find the hardest simplification of the problem that you *can* solve!)

Note that initially in these problems we define the quantity $\binom{n}{r}$ purely algebraically:

$$\binom{n}{r} = \frac{n}{1} \cdot \frac{n-1}{2} \cdot \frac{n-2}{3} \cdots \frac{n-r+1}{r}.$$

No assumption is made that $\binom{n}{r}$ has any combinatorial significance. This will emerge naturally on its own.

★ ★ ★

9.6.1 Study the sequence $\mathbf{A} = \left\langle \binom{n}{2} : n \in \mathbf{N} \right\rangle$:

$$\begin{aligned} \mathbf{A} &= \left\langle \binom{1}{2}, \binom{2}{2}, \binom{3}{2}, \binom{4}{2}, \binom{5}{2}, \binom{6}{2}, \ldots, \right\rangle \\ &= \langle 0,\ 1,\ 3,\ 6,\ 10,\ 15,\ \ldots \rangle, \end{aligned}$$

and prepare a library of its properties.

(The question is open-ended—it invites you to study the given number sequence in the same manner as a chemist would study some newly-manufactured substance. Here are two examples of such properties.

(a) The sequence $D(\mathbf{A})$ formed by the differences between the successive members of $\mathbf{A}$ is

$$\Big\langle 1, 2, 3, 4, 5, 6, \ldots \Big\rangle.$$

This is just the sequence of natural numbers!

(b) If we multiply each member of $\mathbf{A}$ by 8 and add 1 to the result, the resulting sequence (which we denote symbolically by $8\mathbf{A} + 1$) is

$$\Big\langle 1,\ 9,\ 25,\ 49,\ 81,\ 121,\ \ldots \Big\rangle.$$

This is just the sequence of *odd* squares!)

9.6.2 Prove that the number of 2-element subsets of an n-element set is $\binom{n}{2}$.

HINT. Write the subset as $\{x, y\}$ where x, y are distinct. Count the possibilities for x and y separately, then use the fact that $\{x, y\}$ is the same as $\{y, x\}$.

9.6.3 Observe that

$$\binom{2}{2} = 1 = \binom{1}{1};$$
$$\binom{3}{2} = 3 = \binom{1}{1} + \binom{2}{1};$$
$$\binom{4}{2} = 6 = \binom{1}{1} + \binom{2}{1} + \binom{3}{1};$$
$$\binom{5}{2} = 10 = \binom{1}{1} + \binom{2}{1} + \binom{3}{1} + \binom{4}{1}; \quad \ldots.$$

Generalize this sequence of relations; that is, find the general statement for $\binom{n}{2}$, then prove it algebraically. (Use the fact that $\binom{n}{2} = n(n-1)/2$.)

9.6.4 (Continuation) Solve Problem 9.6.3 by exhibiting a one-to-one correspondence between two appropriately chosen sets; that is, a proof in which the quantities on both sides represent the cardinalities of sets.

COMMENT. Such proofs are known as *combinatorial* proofs; we saw a few very pretty examples of such proofs earlier on in the chapter. Combinatorial proofs are typically more elegant than algebraic proofs, which involve only algebraic manipulations and can get rather tedious, but they are more difficult to find!

9.6.5 Prove that the letters of a n-letter word can be arranged in $n!$ distinct ways. (The n letters are assumed to be distinct.)

9.6.6 (Continuation) Prove the natural extension of the above result: *If a n-letter word has a letters of one kind, b letters of another kind, c letters of a third kind, ..., then the number of ways in which its letters can be arranged is*

$$\frac{n!}{a!\,b!\,c!\,\ldots} = \binom{n}{a, b, c, \ldots}.$$

EXAMPLE. The letters of the word PAPA (with $n = 4$, $a = 2$ and $b = 2$) can be rearranged in six ways:

PAPA, PAAP, APPA, APAP, PPAA, AAPP,

and $4!/(2!2!) = 24/4 = 6$. Likewise, the letters of MAMMA (with $n = 5$, $a = 2$ and $b = 3$) can be rearranged in 10 different ways:

MAMMA, MAMAM, MAAMM, MMAMA, MMAAM,
MMMAA, AMMMA, AMMAM, AMAMM, AAMMM,

and $5!/(2!3!) = 120/(2 \times 6) = 10$.

9.6.7 Given an alphabet of n distinct letters, show that the number of r-letter words that one can make without repeating letters is $n!/(n-r)!$. (Here the order of the letters matters; for instance, POST and POTS are different words.)

EXAMPLE. If the alphabet is {A, B, C, D}, the number of 2-letter words is 12. The words are

AB, AC, AD, BA, BC, BD,
CA, CB, CD, DA, DB, DC.

Observe that $4!/(4-2)! = 24/2 = 12$.

9.6.8 (Continuation) What happens if we allow repetitions of letters in the problem posed above? Find a formula for the number of r-letter words that can be made in this case.

EXAMPLE. For $n = 3$ and $r = 2$, the number is 9; e.g., for the alphabet {A, B, C}, the words are

AA, AB, AC, BA, BB, BC, CA, CB, CC.

9.6.9 Prove that a set A with n elements has $\binom{n}{r}$ subsets of cardinality r. (The case $r = 2$ was discussed earlier, in Problem 9.6.2.)

9.6.10 Prove the binomial theorem (for positive integral indices): *For positive integers n, the expansion of $(1+x)^n$ is*

$$1 + \frac{n}{1}x + \frac{n(n-1)}{2!}x^2 + \frac{n(n-1)(n-2)}{3!}x^3 + \cdots + x^n.$$

(The coefficient of x^r is $n!/(r!(n-r)!)$.)

HINT. Write $(1+x)^n$ as $(1+x)(1+x)\cdots(1+x)$, with n bracketed factors in all, then list the ways in which the terms x, x^2, x^3, ... appear in the expansion.

9.6.11 Show that the following identity holds for any positive integer n:

$$\binom{n}{0} + \binom{n}{1} + \binom{n}{2} + \binom{n}{3} + \cdots + \binom{n}{n} = 2^n.$$

There are $n+1$ terms in the sum on the left side.

EXAMPLE. For $n = 3$, we have $1+3+3+1 = 8$.

Look for algebraic as well as combinatorial proofs.

9.6.12 We now know (from Problem 9.6.9) that $\binom{n}{r}$, defined to be the product

$$\frac{n}{1} \cdot \frac{n-1}{2} \cdot \frac{n-2}{3} \cdots \frac{n-r+1}{r},$$

is equal to the number of r-element subsets of an n-element set. Use this fact to find a *combinatorial* proof of the basic recursion relation

$$\binom{n}{r} = \binom{n-1}{r} + \binom{n-1}{r-1}.$$

(The algebraic proof is trivial.)

9.6.13 Use the basic recursion relation $\binom{n}{r} = \binom{n-1}{r} + \binom{n-1}{r-1}$ to find another proof of the assertion that

$$\binom{n}{0} + \binom{n}{1} + \binom{n}{2} + \binom{n}{3} + \cdots + \binom{n}{n} = 2^n.$$

9.6.14 Use the basic recursion relation $\binom{n}{r} = \binom{n-1}{r} + \binom{n-1}{r-1}$ to find another proof of the binomial theorem for positive integers n.

HINT. Write $(1+x)^n$ as $(1+x)\cdot(1+x)^{n-1}$.

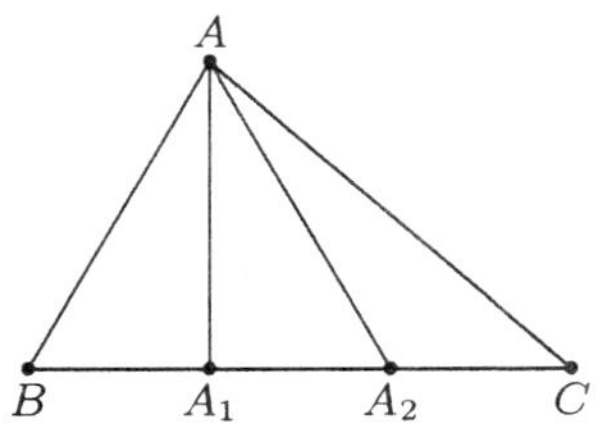

Figure 9.5. *Counting triangles*

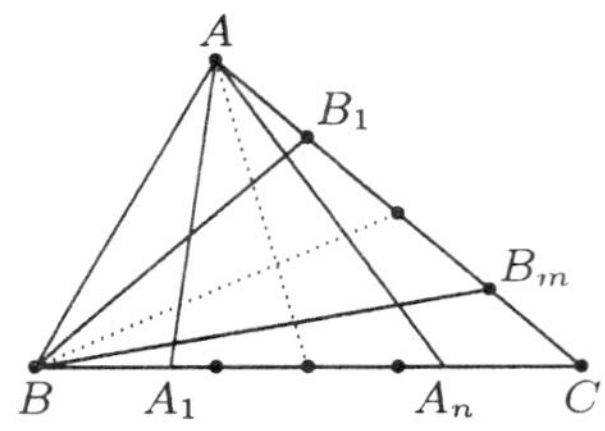

Figure 9.6. *Counting triangles*

9.6.15 Consider all possible subsets of a given finite set A. Let x be the number of subsets with an *odd* number of elements, and let y be the number of subsets with an *even* number of elements. Prove that $x = y$.

EXAMPLE. For $A = \{a, b, c\}$, we find that $x = y = 4$: the 'odd subsets' are $\{a\}$, $\{b\}$, $\{c\}$ and $\{a, b, c\}$, while the 'even subsets' are $\emptyset$, $\{a, b\}$, $\{a, c\}$ and $\{b, c\}$.

Look for both kinds of proofs: algebraic and combinatorial.

9.6.16 In Figure 9.5 we can count as many as 6 triangles. If n distinct points A_1, A_2, A_3, ..., A_n, are marked on side BC of $\triangle ABC$, and the n segments AA_1, AA_2, AA_3, ..., AA_n are drawn, how many triangles may be counted in the figure?

9.6.17 (Continuation) Count the number of triangles in the figure in Figure 9.6.

(There are n distinct points A_1, A_2, ..., A_n on the side BC; m distinct points B_1, B_2, ..., B_m on the side CA and the $m + n$ segments AA_1, AA_2, ..., AA_n, BB_1, BB_2, ..., BB_m are drawn. The answer is to be given in terms of m and n.)

9.6.18 (Continuation) Referring to the figure in the preceding problem, we select k distinct points C_1, C_2, ..., C_k on the side AB, and draw the segments CC_1, CC_2, ..., CC_k. How many triangles are created by this procedure? Find a formula for this number, in terms of n, m, k.

(It may happen that some 3 of the segments (one each through A, B, C) concur, in which case we 'lose' some

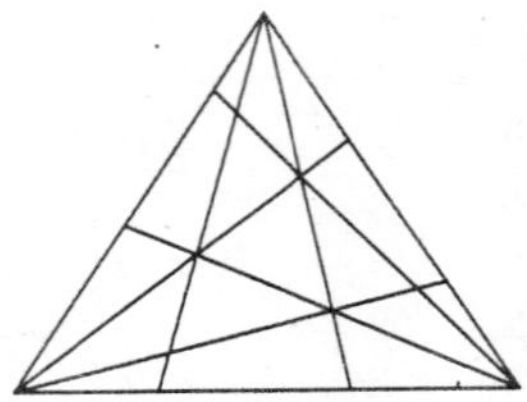

Figure 9.7. *Count the triangles!*

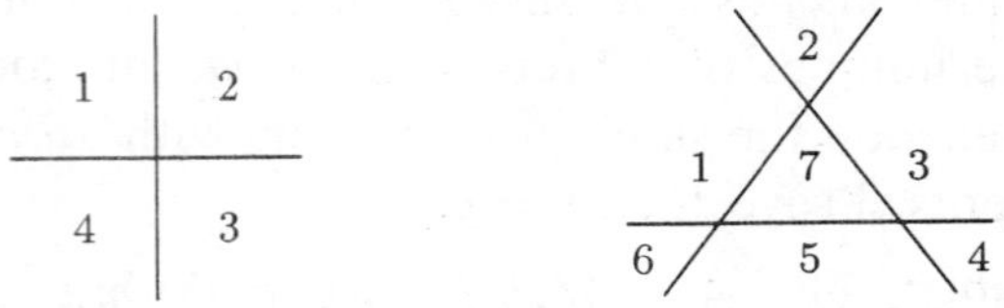

Figure 9.8. *Counting regions*

triangles. Assume that this does not happen. The question now reads: *Find the largest number of triangles that can be obtained when ....*)

9.6.19 (Continuation) If we do allow the concurrence of segments, then some loss occurs in the total count. How do we estimate the loss? Count the number of triangles in Figure 9.7.

9.6.20 When 2 infinitely-long intersecting straight lines are drawn in the plane, 4 disjoint regions are created. And when 3 such lines are drawn, each pair intersecting one another, 7 regions are created as in Figure 9.8.

When n such lines are drawn, at most how many regions are created? If we think of the lines as representing cuts made by a knife, then into how many pieces is the plane cut? Call the number R_n, then $R_1 = 2$, $R_2 = 4$, ... Guess the formula for R_n, then prove it. (As we did earlier, we assume that no three lines pass through one point. So no pieces are 'lost', and the number of pieces is the maximum possible.)

9.6.21 Figure 9.9 shows two triangles divided by grids of parallel

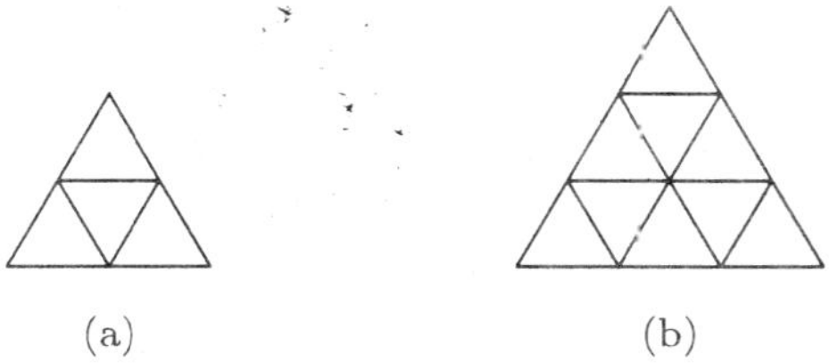

Figure 9.9. *Counting triangles*

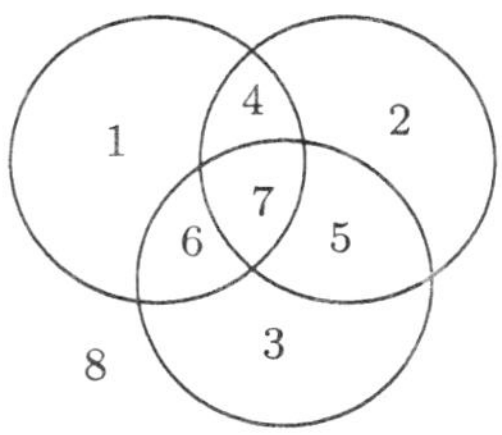

Figure 9.10. *Regions created by* 3 *circles*

lines into smaller triangles, all copies of one another.

Observe that there are $4+1=5$ triangles in (a); they are of two different sizes. In (b) we see a total of $9+3+1=13$ triangles; they are of three different sizes.

Consider the situation when n lines are drawn parallel to each side, spaced out equally. Let T_n be the total number of triangles in the figure ($T_0=1$, $T_1=5$, $T_2=13$, ...). Find a formula for T_n.

Hint. Consider separately the cases n even and n odd.

9.6.22 Let n circles be drawn in the plane, and let C_n be the maximum number of disjoint regions into which the plane can be divided by the circles ($C_1=2$, $C_2=4$, $C_3=8$, ...; the 'outside' counts as a region). Find a formula for C_n.

9.6.23 Let $R_{m,n}$ denote the number of rectangles of all possible sizes in a $m \times n$ rectangular grid ($R_{2,2}=9$, $R_{2,3}=18$, ...). (Note: The term 'rectangle' includes 'square'.) Find a formula for $R_{m,n}$.

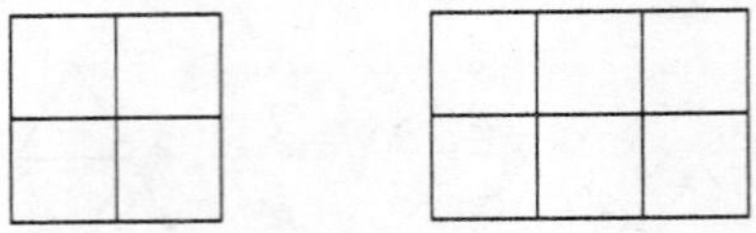

Figure 9.11. *Counting rectangles in a grid*

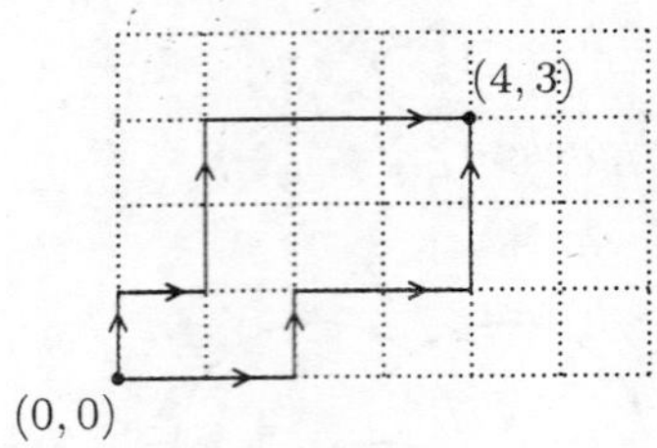

Figure 9.12. *Counting paths in a grid*

9.6.24 (Continuation) Instead of counting rectangles let us now count squares (of all possible sizes). Let $S_{m,n}$ denote this number; then $S_{1,3} = 3$, $S_{2,2} = 5$, ... Find a formula for $S_{m,n}$.

9.6.25 In this problem we count *paths on a grid*. We have before us a semi-infinite rectangular grid of lines, extending upwards and to the right; see Figure 9.12. (A simple model for the grid would be the 1st quadrant of a rectangular co-ordinate system.)

Let co-ordinates be assigned in the natural manner, the bottom left corner being $(0,0)$. Consider the paths joining $(0,0)$ to (m,n) in which: (i) all movements are along the lines of the grid, and (ii) each step is towards the right (East) or upwards (North). Note that there are many such paths connecting two given grid points. (The figure shows 2 paths from $(0,0)$ to $(4,3)$.)

For $m, n \geq 0$, let $P_{m,n}$ denote the number of such paths from $(0,0)$ to (m,n).

EXAMPLE. $P_{2,1} = 3$, $P_{2,2} = 6$.

Find a formula for $P_{m,n}$.

COMMENT. This question arises in taxi-cab geometry, discussed in an earlier chapter.

9.6.26 Observe that the two names AMRISHA and MARISHA are anagrams (rearrangements) of one another. Let all the anagrams of the two names be made, and let them be arranged in alphabetical order (beginning with AAHIMRS, and ending with SRMIHAA). How many words lie in-between AMRISHA and MARISHA?

9.6.27 A positive integer is said to be *sorted* if its digits occur in strictly increasing or strictly decreasing order. Thus, 10, 17, 239, 1458, 54 and 932 are sorted, whereas 254, 255, 655, 768 and 213 are *not* sorted. At the start of this chapter, we asked for the total number of sorted integers, and gave the answer as 1525.

(a) If the list of sorted integers is placed in increasing order, at what position would 4789 occur?

(b) Which would be the 1000th number on the list?

9.6.28 Given a slab of cheese, I plan to cut through it with a knife, making 6 cuts in all, without rearranging the pieces at any of the intermediate stages (the slab is kept intact all through). What is the largest number of cheese-pieces that I can obtain? (The pieces do not have to all be of the same size.)

HINT. First study the 2-dimensional version of the problem: use flat pancakes or omelettes instead of slabs of cheese! Then move to the 3-dimensional version. Use the results from Problem 9.6.20.

9.6.29 Let v_n denote the number of terms in the expansion of the expression $(a+b+c)^n$, after similar terms have been collected together; so $v_1 = 3$, $v_2 = 6$, ... Find a formula for v_n.

9.6.30 (Continuation) Find a formula, in terms of n, for the number of terms in the expansions of the two expressions listed below (as earlier, after like terms have been gathered together).

- $(a+b+c+d)^n$;
- $(a+b+c+d+e)^n$.

Generalize your result.

9.6.31 [*] We are given a 2×4 rectangular sheet of paper divided into unit squares (measuring 1×1), with the numbers 1, 2, ..., 7, 8 written in the squares as shown below.

1	2	3	4
5	6	7	8

The sheet is folded into a 1×1 shape by folding along the grid lines. This causes the 8 numbers to appear within the final compacted piece in some permuted order. (One possible order is $(1, 4, 3, 2, 6, 7, 8, 5)$.)

Clearly, there must exist unachievable permutations. Given an arbitrary permutation, how do we decide whether it is achievable or not? The key question is: *How many permutations are achievable?*

Note that we could have worded the problem with reference to a map measuring 2×4, which is to be folded along its length and width so that the folded-up map measures 1×1.

HINT. Start by examining smaller maps, of sizes 2×2 and 2×3.

9.7 Conclusion

The reader will observe that the level of mathematical maturity required in this chapter is rather more than that required in earlier chapters. As such, one must approach each problem afresh, in a tentative and circumspect manner, so that one's intuition and 'sense' about the problem are gradually built up. Intuition ultimately rests on a base of experiential knowledge and can find expression only after active experimentation and hard work.

⋆ ⋆ ⋆

The contents of this chapter belong to a body of knowledge known as *Enumerative Combinatorics* (more simply, 'The Art of Counting'). We list below a few questions, of a more advanced variety, concerning enumeration.

1. [*] How many sequences $(a_1, a_2, \ldots, a_{2n-1}, a_{2n})$ of length $2n$ exist, in which each a_i is ± 1, $\sum a_i = 0$, and $a_1 + a_2 + \cdots + a_i \geq 0$ for each i?

EXAMPLE. For $n = 2$ there are 2 such sequences, namely $(1, -1, 1, -1)$ and $(1, 1, -1, -1)$.

Let the number be denoted by C_n; then $C_1 = 1$, $C_2 = 2$, $C_3 = 5$, $C_4 = 14$, and so on. These are the *Catalan numbers*. They have a way of cropping up in the most unexpected fashion in problems of enumeration (much like the Fibonacci numbers; see Problem 3, below).

2. [∗] In how many ways can the numbers 1, 2, 3, ..., $2n-1$, $2n$ occupy the $2n$ cells of a $2 \times n$ array, such that, reading from left to right along the rows and from top to bottom down the columns, the numbers occur in strictly increasing order?

3. How many sequences of ones and twos are there whose sum is a given positive integer n? (The order of the summands matters, but the length of the sequence is not fixed.

EXAMPLE. For $n = 4$ the relevant sequences may be written as: 22, 211, 121, 112 and 1111; there are 5 such sequences.)

Let the number be denoted by F_n; then $F_1 = 1$, $F_2 = 2$, $F_3 = 3$, $F_4 = 5$, ... These are the *Fibonacci numbers*. They are even more interesting than the Catalan numbers!

4. [∗] In how many ways can a positive integer n be expressed as a sum of non-negative integers (with order not regarded as important)?

EXAMPLE. For $n = 4$ the possibilities are 4, $3+1$, $2+2$, $2+1+1$ and $1+1+1+1$; there are 5 possible ways.

Let the number be denoted by $p(n)$; then $p(1) = 1$, $p(2) = 2$, $p(3) = 3$, $p(4) = 5$, $p(5) = 6$, and so on. The function p is known as the *partition function*, and the mathematics associated with it is very deep and beautiful. Ramanujan and Hardy have to their name an extremely surprising theorem about $p(n)$.

5. The numbers 1, 2, 3, ..., $n-1$, n are to be written in the cells of a $n \times n$ array, each number being repeated n times, so that no number repeats in any row or column. (This means that each row contains n different numbers,

and likewise for each column.) Such arrays are called *Latin squares*, and there is a rich body of theory behind them. The question is: *How many Latin squares are there for any given n?*

6. [*] In how many distinguishable ways can the faces of a cube be coloured using the colours Red, Black and White? This question is studied in *Polya enumeration theory*. An understanding of symmetry and group theory is needed to study this topic, for the counting must be done with reference to the group of symmetries of the object.

⋆ ⋆ ⋆

The number of directions of possible further study is clearly very large. Enumerative techniques sometimes have unexpected applications in other disciplines. In part, this is because the principles of counting are so basic in nature that they are potentially of use just about anywhere. For instance, there are applications in Organic Chemistry (enumeration of the isomers that share a given chemical formula); Genetics (the mathematics behind the genetic code); Statistics (the design of experiments in Agriculture and Psychology; e.g., the testing of newly-patented drugs); Physics (Maxwell–Boltzman statistics, Fermi–Dirac statistics, Bose–Einstein statistics. Here, the number of ways in which the particles of a gas can occupy a given set of available energy levels is counted, and the count is taken to be a measure of the likelihood of finding the gas in some particular state. The rules governing the occupancy of the different energy levels differ for the three types of statistics); Economics is yet another field; and so on.

⋆ ⋆ ⋆

In conclusion, we offer a few non-enumerative problems.

1. (This problem looks hard but can be solved very simply, using a few basic principles.)

   Let a string σ of digits be given of any finite length k, however large:

$$\sigma = (a_1, a_2, a_3, \ldots, a_{k-1}, a_k).$$

   (Each a_i is a digit.) Show that there exists a positive integer N such that the decimal part of $\sqrt{N}$ starts off *precisely* with the string σ. That is, the decimal expansion of $\sqrt{N}$ has the following appearance:

$$\sqrt{N} = I.a_1\, a_2\, a_3\, \ldots\, a_{k-1}\, a_k\, a_{k+1}\, a_{k+2} \ldots,$$

where I is an integer and $a_{k+1}, a_{k+2}, \ldots$ are digits.

A more satisfactory formulation of the problem is as follows. Let ϵ be a number between 0 and 1 and let k be any positive integer, however large. Then there exist positive integers N and I such that

$$0 < \sqrt{N} - I - \epsilon < 10^{-k}.$$

EXAMPLE. Let $k = 4$, $\sigma = (7, 1, 6, 2, 3)$. After some searching, we find that $N = 247$ "works" ($\sqrt{247} = 15.71623\ldots$).

2. (Two related problems from Paul Erdös; the first one was once posed by him to a young colleague, Louis Posa, then 14 years old; he solved it over breakfast!)

   (a) Let $n + 1$ numbers be chosen from $\{1, 2, 3, \ldots, 2n\}$. Show that among the chosen numbers there exist two that are coprime (that is, they share no common factor). Show further that this conclusion is false if $n + 1$ is replaced by n.

   (b) Let $n + 1$ numbers be chosen from $\{1, 2, 3, \ldots, 2n\}$. Show that among the chosen numbers there exist two, say a and b with $a < b$, such that $a \mid b$ (that is, a is a divisor of b). Show further that this conclusion is false if $n + 1$ is replaced by n.

3. (Continuation) Let $3n + 1$ numbers be chosen from the set $\{1, 2, 3, \ldots, 4n\}$. Show that among the chosen numbers there exist three, say a, b and c with $a < b < c$, such that $a \mid b$ and $b \mid c$. Show further that this conclusion is false if $3n + 1$ is replaced by $3n$.

4. Let any ten 2-digit numbers be chosen. Show that there exist two disjoint subsets of these numbers that have equal sum. (This is an IMO problem.)

5. Let any 5 points be selected in a square measuring 2×2. Show that there are 2 points among them which are at a distance of at most $\sqrt{2}$ from one another.

6. Let α be any irrational number and ϵ any positive real number, however small. Show that there exists an integer n such that the decimal part of $n\alpha$ is less than ϵ (that is, $0 < n\alpha - [n\alpha] < \epsilon$).

7. Seven polygons of area 1 lie within the interior of a square of side 2. Prove that at least two of the polygons intersect in a region of area $\geq 1/7$.

8. Let S be the set $\{1, 2, \ldots, 2n - 1, 2n\}$. Find the size of the largest sum-free subset of S. (A set is 'sum-free' if there do not exist three numbers a, b, c in it, not necessarily distinct, such that $a + b = c$. Thus, $\{1, 3, 7\}$ is sum-free, whereas $\{1, 2, 4\}$, $\{1, 3, 6\}$ and $\{2, 3, 5\}$ are not sum-free.)

9. Let A be any positive 16-digit integer. Show that we can find a consecutive block of digits of A whose product is a square.

   Is the result true for 15-digits numbers too?

10. Determine the least number n with the following property: for every partition of the set $\{3, 4, 5, \ldots, n\}$ into two disjoint subsets, at least one of the subsets contains numbers a, b, c, not necessarily distinct from one another, such that $ab = c$.

We have listed these problems together as they are all solved by using the *pigeon-hole principle* (PHP for short). It sounds very innocuous, but is in fact extremely powerful:

> *If* $n+1$ *pigeons occupy* n *pigeon-holes, then at least one pigeon-hole will contain 2 or more pigeons.*

The task of identifying the pigeons and pigeon-holes in the problems listed above is left to the reader.

9.8 References

A rich offering is available in the existing literature for the student wishing to go deeper into the study of combinatorics. The following list includes just a few suggestions.

[1] Barnard & Child. *Higher Algebra* (Macmillan).

[2] Cohen, Daniel. *Combinatorics.*

[3] Niven, Ivan. *The Art of Counting* (MAA).

[4] Sved, Marta. *Counting and Recounting*, in *The Mathematical Intelligencer*, Volume **5**, Number 4 (1983), pg 21–26.

[5] Vilenkin, Y. *Combinatorics.*

[6] Tomescu, Ioan & Melter, Robert. *Problems in Combinatorics and Graph Theory.*

References [1], [2] and [6] are highly recommended; [1] has a chapter on Combinatorics, and [6] has an extremely rich problem collection. [4] is a delightful article, focussing on proofs of binomial identities using combinatorial (and non-computational) arguments.

Chapter 10

Number Theory

10.1 Introduction

The subject matter of this chapter has always been popular with amateurs, perhaps because the concepts used are so elementary and there is so much material to study that the subject can serve as an endless source of pleasure. Take, for instance, the notion of primality of numbers, which we encounter while quite young. Questions concerning prime numbers are studied at the highest level even today, because there are so many unanswered questions in the field. For instance, consider the following questions.

(a) Are there infinitely many instances of *twin primes* (primes that differ by 2)? For example, the pairs

$$(3,5),\ (5,7),\ (11,13),\ (17,19),$$
$$(29,31),\ (41,43),\ (71,73),\ (101,103)$$

provide instances of twin primes.

(b) Is there a prime number between n^2 and $(n+1)^2$ for every positive integer n?

EXAMPLE. There are primes between 10^2 and 11^2 (e.g., the prime 101), between 50^2 and 51^2 (e.g., the prime 2503), and between 10000^2 and 10001^2 (e.g., the prime 100000007).

(c) Are there infinitely many primes of the form $2^n - 1$?

EXAMPLE. Primes of the stated form include $2^3 - 1 = 7$, $2^5 - 1 = 31$, and $2^{13} - 1 = 8191$. There are many more!

No-one knows the answers to these questions! (The available evidence suggests that each answer is *yes*.)

Or take the notion of perfection. A positive integer exceeding 1 is said to be *perfect* if it equals the sum of its proper divisors. Examples include 6 and 28 ($1+2+3=6$, $1+2+4+7+14=28$). Consider the following two questions:

(d) Are there infinitely many perfect numbers?

(e) Are there any odd perfect numbers?

Once again, no one knows the answers! (The likely answers are *yes* and *no* respectively, but no one really knows.) We quote at this point two related theorems on perfect numbers, the first from Euclid's famous text, *The Elements* (we have rewritten it in modern language), and the second from Euler.

- If P is a prime number of the form $2^n - 1$ for some positive integer n, then the number $P(P+1)/2$ is perfect.
- Every even perfect number is of the form $P(P+1)/2$, where P is a prime number of the form $2^n - 1$ for some positive integer n.

Since 3 and 7 are primes of the stated form (each is 1 less than a power of 2), according to Euclid, the numbers $\frac{1}{2}(3 \times 4) = 6$ and $\frac{1}{2}(7 \times 8) = 28$ ought to be perfect; and this is so. The next such prime is 31, and $\frac{1}{2}(31 \times 32) = 496$ is perfect. Note that the two results, when taken together, provide a complete characterization of perfect even numbers. So, between them, Euclid and Euler have said the last word on this class of numbers! (Small consolation: the list of such numbers known to us today is much larger than theirs!)

Concerning odd perfect numbers, the facts known about them are all of recent origin. In brief, they assert that such numbers, if they exist at all, must be exceedingly large.

⋆ ⋆ ⋆

Number theory is essentially concerned with topics such as these: properties of the primes; the divisors of integers; the solutions of equations in integers; and so on. The principles involved are so elementary – mostly of the 'utterly obvious' variety – that it is astonishing how far they can take us. It is like taking a journey into vast and distant lands, carrying almost nothing at all.

10.2 Proof by contradiction

One thing that does need to be taken along is the notion of proof by contradiction (known in Latin as *reductio ad absurdum*). We have already dwelt on this in Chapter 3, where some instances were presented of proof by contradiction in geometry. (It often comes of use in proving converses.) But it is in number theory and combinatorics where the idea is seen at its dramatic best.

The idea rests on the following fundamental axiom:

Inconsistency or self-contradiction cannot be accepted.

This offers us the following strategy for proving statements that we feel are true.

Let $\mathcal{P}$ be the statement (*proposition*, to use the technical term) we wish to prove. We start by assuming that the proposition is false; that is, we take the *negation* of $\mathcal{P}$, which we denote by $\sim\mathcal{P}$, to be true. The consequences of this assumption are now pursued relentlessly. If at some point two consequences are found that contradict one another, then the desired inconsistency or self-contradiction has been found. The implication now is that the assumption made at the start (namely, $\sim\mathcal{P}$) must be false. Since a double negation gets us back to the original proposition, $\mathcal{P}$ must be true!

This is essentially the way in which all proofs of this kind are organized. Today, the idea of proof by contradiction constitutes a vital weapon in the armory of every mathematician. We give below a few instances of this principle at work:

Question *Is there a largest integer?*

Suppose there is a largest integer, N. Let $\mathcal{P}$ be the proposition, *"N is the largest integer"*. Now if N is an integer, so is $N+1$; and $N+1$ exceeds N. So we have found a larger integer. Thus, $\mathcal{P}$ is self-contradictory, and therefore false. It follows that there can be no largest integer. □

One can similarly dispose of questions such as: *"Is there a largest square number?"* and *"Is there a largest composite number?"*. (Both yield to the following observation: for any integer $N > 1$, N^2 is square, it is composite, and it exceeds N. Please fill in the details on your own.)

These may seem silly questions to have started with; few people would have any doubts concerning their answers! However, the next three examples are anything but trivial.

Infinitude of the primes We ask the following question, and then give Euclid's beautiful answer.

Is there a largest prime number?

Let $\mathcal{A}$ denote the following proposition: *"There exists a largest prime number P"*. We shall suppose that $\mathcal{A}$ is true. Let N be the number defined as follows:

$$N = (2 \times 3 \times 5 \times 7 \times \cdots \times P) + 1.$$

That is, N is 1 plus the product of all available primes. We now ask: *Is N prime or composite?*

Since N exceeds P, by invoking the hypothesis $\mathcal{A}$, we infer that N is composite (for, according to $\mathcal{A}$, there are no primes exceeding P). Therefore, N has prime factors. However, N leaves a remainder of 1 when divided by each of the primes 2, 3, 5, 7, ..., P, so N is indivisible by each of them. By supposition, these are *all* the primes that exist, so we conclude that N is not divisible by any prime number. We now have an absurd situation: N is not prime, not a unit (it certainly exceeds 1), yet it is indivisible by all the primes! Of course, this is a self-contradiction and therefore unacceptable, so we conclude that $\mathcal{A}$ is false. In other words, that there does not exist any largest prime number. So the number of primes is infinite. □

⋆ ⋆ ⋆

The reader may wish to do some number-crunching here. Write p_n for the nth prime and $f(n)$ for the number $(2\times3\times5\times\cdots\times p_n)+1$. A few values of $f(n)$ are given in the table below.

n	1	2	3	4	5	6	7	...
p_n	2	3	5	7	11	13	17	...
$f(n)$	3	7	31	211	2311	30031	510511	...

We find that that 3, 7, 31, 211 and 2311 are prime; whereas, 30031 and 510511 are composite:

$$30031 = 59 \times 509, \qquad 510511 = 19 \times 97 \times 277.$$

Observe that the prime factors of 30031 exceed 13, and the prime factors of 510511 exceed 17. This agrees with what Euclid had asserted.

★ ★ ★

One would be hard put to find a finer example of proof by contradiction; the proof is a classic, a gem.[1] The very quotable words of G H Hardy are brought to mind by this example:

> In a chess game a player sometimes offers a pawn as a gambit, with the intention of winning it back (or more) later in the game. But in proof by contradiction, the mathematician offers the whole theorem—the whole game itself, as it were ....

★ ★ ★

COMMENT. In passing, we note that there is an unsolved problem connected with the above sequence.

The question is: *Is $f(n)$ prime for infinitely many n?* We find that $f(1)$, $f(2)$, $f(3)$, $f(4)$ and $f(5)$ are prime. Which is the next one? We have the following data:

$$\begin{aligned}
f(6) &= 30031 = 59 \times 509 \\
f(7) &= 510511 = 19 \times 97 \times 277 \\
f(8) &= (2 \times 3 \times 5 \times \cdots \times 17 \times 19) + 1 \\
&= 9699691 = 347 \times 27953 \\
f(9) &= (2 \times 3 \times 5 \times \cdots \times 19 \times 23) + 1 \\
&= 223092871 = 317 \times 703763 \\
f(10) &= (2 \times 3 \times 5 \times \cdots \times 23 \times 29) + 1 \\
&= 6469693231 = 331 \times 571 \times 34231 \\
f(11) &= (2 \times 3 \times 5 \times \cdots \times 29 \times 31) + 1 \\
&= 200560490131,
\end{aligned}$$

which is prime; so the next prime after $f(5)$ is $f(11)$. The numbers grow rapidly and soon become unmanageable. In general, what can be said about the sequence $\langle f(n) \rangle$? No-one knows whether it contains infinitely many prime numbers.

---

[1] Note, however, that the proof can be written in a constructive style, avoiding the idea of self-contradiction.

Irrationality of the square root of 2 Our next example, another classic, again goes back to Euclid. We ask the following question:

> *Is $\sqrt{2}$ a rational number? That is, is there a rational number whose square equals 2?*

We suppose that there is such a rational number, a/b, where a, b are positive integers with no common factor. This supposition leads to the following consequences:

(1) $\dfrac{a^2}{b^2} = 2, \quad \therefore\ a^2 = 2b^2$, an even number.

(2) a is even (for the square of an odd integer is odd), so $\frac{1}{2}a$ is an integer, say $\frac{1}{2}a = c$.

(3) $a = 2c, \quad \therefore\ a^2 = 4c^2, \quad \therefore\ 2b^2 = 4c^2.$

(4) $b^2 = 2c^2$, an even number, $\therefore\ b^2$ is even, $\therefore\ b$ itself is even.

(5) Therefore, both a and b are even.

We had started out by supposing that a, b had no common factor and now we find that they share the factor 2; we have a self-contradiction! So we must drop our original supposition, and we conclude that $\sqrt{2}$ is irrational. □

Divergence of the harmonic series Our next example is associated with one of the great names of 16th century mathematics—Daniel Bernoulli. We ask the following question:

> *Is the sum $1 + \frac{1}{2} + \frac{1}{3} + \frac{1}{4} + \cdots$ (to infinity) a finite quantity?*

This question is motivated by the observation that it *is* possible to add infinitely many positive numbers and yet obtain a finite answer. For example, the sum

$$\frac{1}{10} + \frac{1}{100} + \frac{1}{1000} + \frac{1}{10000} + \frac{1}{100000} + \cdots \text{ (to infinity)}$$

is finite. (The sum corresponds to the infinite recurring decimal $0.11111\ldots$, which equals the fraction $1/9$.)

The reader may complain, quite justifiably, that it does not make sense to 'add' infinitely many numbers. How could we do such a thing?—i[illegible]ld take an infinite amount of time! If we want to be more prudent and precise in our usage of language – a good practice in general! – then we must write: *The limiting value of the sums* .1, .1 + .01, .1 + .01 + .001, .1 + .01 + .001 + .0001, . . . *is* 1/9, or simply: *The limiting value of the sum* .1 + .01 + .001 + ⋯ *is* 1/9, rather than *"The sum of the quantities . . . is* 1/9*"*. The chief difference between the two modes of expression lies in using the phrase 'limiting value of the sum' instead of 'sum'.

When a sum of infinitely many quantities can be meaningfully computed in this manner, that is, as a limit of sums of finite numbers of quantities, the sum is said to *converge.*

EXAMPLE. The following two series:

$$\frac{1}{2}+\frac{1}{4}+\frac{1}{8}+\frac{1}{16}+\frac{1}{32}+\frac{1}{64}+\cdots \text{ (to infinity)},$$
$$\frac{1}{2}-\frac{1}{4}+\frac{1}{8}-\frac{1}{16}+\frac{1}{32}-\frac{1}{64}+\cdots \text{ (to infinity)},$$

converge, to 1 and 1/3, respectively. Of course, not all infinite series converge. For instance, the following series:

$$1+1+1+1+\cdots \text{ (to infinity)},$$
$$1+2+3+4+\cdots \text{ (to infinity)},$$
$$1-1+1-1+\cdots \text{ (to infinity)},$$

very obviously fail to converge.

The question posed above may be restated as follows.

Determine whether or not the following sum converges:

$$\frac{1}{1}+\frac{1}{2}+\frac{1}{3}+\frac{1}{4}+\cdots.$$

The obvious answer, 'yes', is incorrect. This may seem surprising, because the numbers 1/1, 1/2, 1/3, 1/4, . . . form a steadily decreasing sequence, and it would appear that their sum ought to converge to a finite limit. However, this is not the case. Here is one of several proofs of the result, due to Daniel Bernoulli. (The result was first proved in the 1600s by Nicolo Oresme.)

Assume that there is a limiting value L. We write:

$$L = \frac{1}{1}+\frac{1}{2}+\frac{1}{3}+\frac{1}{4}+\frac{1}{5}+\frac{1}{6}+\cdots,$$
$$\therefore \frac{L}{2} = \frac{1}{2}+\frac{1}{4}+\frac{1}{6}+\frac{1}{8}+\frac{1}{10}+\frac{1}{12}+\cdots.$$

Subtracting the second equation from the first, we obtain

$$\frac{L}{2} = \frac{1}{1} + \frac{1}{3} + \frac{1}{5} + \frac{1}{7} + \frac{1}{9} + \frac{1}{11} + \cdots.$$

Therefore, we must have

$$\frac{1}{1} + \frac{1}{3} + \frac{1}{5} + \cdots = \frac{1}{2} + \frac{1}{4} + \frac{1}{6} + \cdots,$$

which is absurd, because

$$\frac{1}{1} > \frac{1}{2}, \quad \frac{1}{3} > \frac{1}{4}, \quad \frac{1}{5} > \frac{1}{6}, \quad \cdots.$$

We have arrived at a self-contradiction. Thus, there is no limiting value L to the given sum!

Historically, this was the first non-trivial instance of a sum being shown to be infinite.

COMMENT. There are other ways of writing the proof, which are more "constructive" in nature. The simplest of these goes roughly as follows. Let $S(n)$ denote the sum $1 + 1/2 + \cdots + 1/n$; i.e., the sum of the first n terms. Then it is possible to show quite easily that $S(2^n) > n/2$. So, by choosing n to be large enough, $S(2^n)$ can be made to exceed any given quantity. This means that $S(n)$ does not converge to any finite limit; i.e., the sum does not converge.

⋆ ⋆ ⋆

This, then, is how proof by contradiction works. It is a powerful weapon, but be forewarned that it is tricky to use. Some teachers actively discourage their students from using it at too early an age, because of the observed tendency of many students to rashly conclude a line of argument with an invalid self-contradiction. An additional source of difficulty, from a teacher's point of view, is that often the self-contradictory element bears little relationship to the original claim. This often makes even valid proofs by contradiction seem rather obscure at times.

In the problem sets that follow, proof by contradiction is to be used wherever appropriate.

10.3 Preliminaries

10.3.1 Notation

Before proceeding further, we shall introduce some important symbols. We will find that they are of great benefit in presenting arguments in a clean and succinct form. Indeed, they also help us *think* in a more precise manner.

- **REMAINDERS** Let a, b be integers, with b non-zero. When we divide a by b, we denote the (non-negative) remainder by $\text{Rem}(a \div b)$. So $\text{Rem}(a \div b)$ is the least non-negative integer x such that $a - x$ is divisible by b.

  EXAMPLE. We have $\text{Rem}(21 \div 6) = 3$, $\text{Rem}(100 \div 7) = 2$ and $\text{Rem}(-50 \div 11) = 5$.

- **DIVISIBILITY** Let a, b be integers, with a non-zero. If a is a divisor of b, we write $a \mid b$; and if a is not a divisor of b, then we write $a \nmid b$. Observe that '$a \mid b$' can be stated alternatively as: 'b/a is an integer'.

  EXAMPLE. We have $3 \mid 9$ and $7 \mid 1001$, but $5 \nmid 21$ and $11 \nmid 101$. Also, $a \mid 0$ for all $a \neq 0$.

- **GREATEST COMMON DIVISOR** Let a and b be integers, not both zero. The greatest common divisor of a and b, written as $\gcd(a, b)$, is the largest positive integer d such that $d \mid a$ and $d \mid b$. Note that $\gcd(0, 0)$ is left undefined.

  If $\gcd(a, b) = 1$, then we say that the integers a and b are *relatively prime*, or that they are *coprime*.

  EXAMPLE. We have $\gcd(10, 15) = 5$ and $\gcd(14, 21) = 7$. Also, $\gcd(a, 0) = a$ for all non-zero integers a, and $\gcd(a, a+1) = 1$ for all positive integers a.

  The following laws which hold for the last two symbols are easy to prove:

  (i) If $a \mid b$ and $b \mid c$, then $a \mid c$.

  (ii) If $d \mid a$ and $d \mid b$, then $d \mid a+b$, $d \mid a-b$; and more generally, $d \mid ma + nb$ for all integers m and n.

  (iii) $\gcd(a, b) = \gcd(a, a+mb) = \gcd(a+nb, b)$ for all integers m and n.

- **CONGRUENCES** Let a, b, c be integers, with c being non-zero. We shall write

  $$a \equiv b \pmod{c},$$

  to mean that $\text{Rem}(a \div c) = \text{Rem}(b \div c)$. That is, a and b leave the same remainder when divided by c; equivalently, $c \mid a - b$.

The statement $a \equiv b \pmod{c}$ is to be read aloud as "a is congruent to b modulo c" (the "modulus" is c).

EXAMPLE. We have $7 \equiv 2 \pmod 5$, $9 \equiv 1 \pmod 8$, $39 \equiv 18 \pmod 7$ and $1000 \equiv -1 \pmod{11}$.

The congruence symbol is very widely used, and its usefulness is due to the fact that one can use it more or less like one uses the 'equals to' symbol, '='; the same laws apply for both the symbols (well, almost). So, given that $a \equiv b \pmod k$ and $c \equiv d \pmod k$ (observe that the modulus is the same in both relations), we may deduce each of the following (all with reference to the same modulus, k):

$$a + c \equiv b + d, \quad ac \equiv bd, \quad a^n \equiv b^n, \text{ for all positive integers } n;$$

and so on. Many important results about integers may be stated in a compact manner and deduced with great ease with the use of this symbol. The following statements, all of which refer to an arbitrary integer n, provide good instances of this claim.

1. For all n we have $n^2 \equiv 0$ or $1 \pmod 3$; we can never have $n^2 \equiv 2 \pmod 3$.

   EXAMPLE. We have $5^2 \equiv 1 \pmod 3$, $6^2 \equiv 0 \pmod 3$, and $7^2 \equiv 1 \pmod 3$.

2. For all n we have $n^2 \equiv 0$ or $1 \pmod 4$; we can never have $n^2 \equiv 2$ or $3 \pmod 4$.

   EXAMPLE. We have $5^2 = 25 \equiv 1 \pmod 4$, $6^2 = 36 \equiv 0 \pmod 4$, and $7^2 = 49 \equiv 1 \pmod 4$.

3. If n is odd, then $n^2 \equiv 1 \pmod 8$.

   EXAMPLE. We have $5^2 = 25 \equiv 1 \pmod 8$ and $11^2 = 121 \equiv 1 \pmod 8$.

4. For odd n we have $2^n + 1 \equiv 0 \pmod 3$ (equivalently, $3 \mid 2^n + 1$ for all odd n). Conversely, $3 \nmid 2^n + 1$ if n is even.

   EXAMPLE. We have $2^3 + 1 = 9 \equiv 0 \pmod 3$, $2^5 + 1 = 33 \equiv 0 \pmod 3$, and $2^6 + 1 = 64 \not\equiv 0 \pmod 3$.

5. The relation $4 \mid 3^n + 1$ is true for odd n and false for even n.

   EXAMPLE. We have $4 \mid 28 = 3^3 + 1$ and $4 \nmid 82 = 3^4 + 1$.

10.3.2 Simple applications

We give below three simple instances of how congruences are used in proving statements in number theory. First, we shall state the problems.

(1) Given a list of 1998 odd numbers, is it possible that the square of one of them is equal to the sum of the squares of the other 1997 numbers?

(2) Show that $2^n + 1$ is divisible by 3 whenever n is odd, but not when n is even. (This is item #4, above.)

(3) Show why the familiar test for divisibility by 9 works. (The test is for numbers in base-10. It states that a number is divisible by 9 if and only if the sum of its digits is divisible by 9.)

Solution to problem 1 Let X be the sum of the squares of any 1997 odd numbers; note that X is odd. We now use item #3 of the preceding section: if n is odd, then $n^2 \equiv 1 \pmod 8$. From this, we obtain $X \equiv 1997 \pmod 8$. Since $1997 \equiv 5 \pmod 8$, we obtain $X \equiv 5 \pmod 8$. But no odd square is congruent to 5 (mod 8), so X cannot be a square.

Solution to problem 2 Since $2 \equiv -1 \pmod 3$, we get

$$2^n + 1 \equiv (-1)^n + 1 \pmod 3.$$

But $(-1)^n = \pm 1$; the minus sign holds if n is odd, and the plus sign if n is even. So $(-1)^n + 1$ is 0 or 2, depending only on whether n is odd or even. It follows that $2^n + 1 \equiv 0 \pmod 3$ when n is odd, and $2^n + 1 \equiv 2 \pmod 3$ when n is even.

Solution to problem 3 Let the decimal digits of n, reading from left to right, be a, b, c, ..., and let the number of digits be k; then we have

$$n = a \cdot 10^{k-1} + b \cdot 10^{k-2} + c \cdot 10^{k-3} + \cdots.$$

From $10 \equiv 1 \pmod 9$ we get $10^i \equiv 1 \pmod 9$ for all non-negative integers i, and therefore $n \equiv a + b + c + \cdots \pmod 9$. So if Sum($n$)

denotes the sum of the digits of n, then

$$n \equiv \text{Sum}(n) \pmod 9,$$

i.e., $9 \mid n$ if and only if $9 \mid \text{Sum}(n)$.

10.4 Pythagorean triples

An excellent illustration of the methods used in elementary number theory is seen in the study of Pythagorean triples.

Pythagoras' theorem is well-known: *If $\triangle ABC$ is right-angled at vertex C, then its sides a, b, c obey the relation $a^2 + b^2 = c^2$.* The converse statement is true too: *If a, b, c are positive numbers such that $a^2 + b^2 = c^2$, then a, b, c are the sides of a right-angled triangle, say $\triangle ABC$, right-angled at C.*

If $\triangle ABC$ is right-angled at vertex C, and if the sides a, b, c are integers, then the triple (a, b, c) is referred to as a *Pythagorean triple*, or PT for short. If in addition the integers a, b, c are coprime, then we refer to the triple as a *primitive Pythagorean triple*, or PPT for short.

EXAMPLE. The triple $(3, 4, 5)$ is a PPT, perhaps the best-known one! Other examples are $(5, 12, 13)$ and $(8, 15, 17)$. To construct a PT which is not a PPT, we simply "scale up" a PPT; examples: $(6, 8, 10)$ and $(10, 24, 26)$.

The main result that we shall prove concerning PPTs is the following.

Theorem 1 *Let (a, b, c) be a PPT, and suppose that b is odd. Then there exist odd positive integers p and q with no common factors, with $p > q$, such that*

$$a = \frac{p^2 - q^2}{2}, \qquad b = pq, \qquad c = \frac{p^2 + q^2}{2}.$$

NOTE. Since (a, b, c) is a PPT, a and b cannot both be even. At least one of them is odd and we shall suppose, without any loss that b is odd. (No supposition has been made concerning the parity of a. In fact, a must be even; indeed, a multiple of 4.)

EXAMPLE. The PPT $(4, 3, 5)$ corresponds to $p = 3$, $q = 1$, while $(12, 5, 13)$ corresponds to $p = 5$, $q = 1$.

Proof We write the relation as $b^2 = c^2 - a^2 = (c-a)(c+a)$. Since b by supposition is odd, it follows that b^2 is odd and so $c-a$ and $c+a$ are both odd. (The divisors of an odd number are odd.) What might be the greatest common divisor (gcd) of $c-a$ and $c+a$? Let the gcd be d, then d is odd since $c-a$ and $c+a$ are both odd. Next, d must divide the sum and difference of the two quantities $c-a, c+a$ (because it divides each of them separately); so we get $d \mid 2c$ and $d \mid 2a$. We have just observed that d is odd, so the '2' in these two relations can be dropped, and we get $d \mid c$, $d \mid a$; $\therefore d$ is a common divisor of c and a. But c, a are coprime (for (a, b, c) is a *primitive* triple), $\therefore d = 1$; that is, the quantities $c-a$ and $c+a$ are coprime. Since their product is a square, it follows that each of them is a square. Let $c+a = p^2$ and $c-a = q^2$, where p and q are (odd) positive integers with a gcd equal to 1. This yields:

$$a = \frac{p^2 - q^2}{2}, \quad b = pq, \quad c = \frac{p^2 + q^2}{2},$$

as claimed. □

The formula allows us to generate infinitely many PPTs. For example,

- from $p = 7$ and $q = 1$ we get the PPT $(24, 7, 25)$;
- from $p = 5$ and $q = 3$ we get the PPT $(8, 15, 17)$;
- from $p = 7$ and $q = 3$ we get the PPT $(20, 21, 29)$;

and so on.

Theorem 1 may be stated in another form which may be found to be more convenient:

> *If (a, b, c) is a* PPT *and b is odd, then there exist coprime integers u and v of opposite parity, with $u > v$, such that*
>
> $$a = 2uv, \quad b = u^2 - v^2, \quad c = u^2 + v^2.$$

Here, 'opposite parity' means that one of u and v is even and the other is odd, and "coprime" means that the two integers have no factors in common.

EXAMPLE. From $u = 4$, $v = 1$ we get the PPT $(8, 15, 17)$.

The proof of equivalence of the two generating formulas is left to the reader. We now present another proof of the result (in the

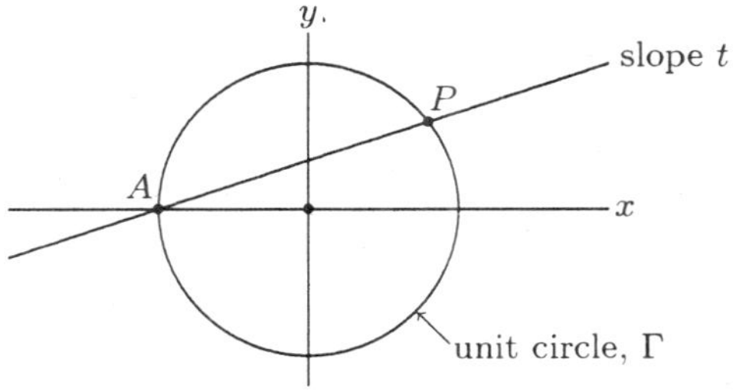

Figure 10.1. *Enumerating rational points on the unit circle*

second form). It proceeds in a totally different way, combining algebra with geometry.

Let (a, b, c) be a PPT, and let $x = a/c$, $y = b/c$; then x and y are rational numbers, and $x^2 + y^2 = 1$, so the pair (x, y) corresponds in the coordinate plane to a rational point $P(x, y)$ on the circle Γ with unit radius and center $O = (0, 0)$ (a 'rational point' is a point for which both coordinates are rational). So, in effect, we seek a scheme for generating all the rational points on the unit circle (see Figure 10.1).

Consider the point $A = (-1, 0)$ on Γ. For any rational number t, consider the line with slope t passing through A; its equation is $y = t(x+1)$. Let the line intersect Γ again at the point $P_t = (x_t, y_t)$. To find x_t and y_t we must solve the equations $y = t(x + 1)$, $x^2 + y^2 = 1$. Eliminating y, we obtain the quadratic equation

$$(t^2 + 1)x^2 + 2t^2x + (t^2 - 1) = 0.$$

This equation has $x = -1$ as a root (check this for yourself!), corresponding to the point A itself. To find the other root, we factorize the LHS of this equation, using the fact that we already know one factor (i.e., $x + 1$). We find that

$$(t^2 + 1)x^2 + 2t^2x + (t^2 - 1) = (x + 1)\left((t^2 + 1)x + (t^2 - 1)\right).$$

The root x_t is obtained by solving $(t^2 + 1)x + (t^2 - 1) = 0$; we get

$$x_t = -\frac{t^2 - 1}{t^2 + 1} = \frac{1 - t^2}{1 + t^2}.$$

The value of y corresponding to this is found using the equation $y = t(x + 1)$; we get $y_t = t(1 + (1 - t^2)/(1 + t^2) = 2t/(1 + t^2)$. So we have the coordinates of P_t:

$$P_t = \left(\frac{1 - t^2}{1 + t^2}, \frac{2t}{1 + t^2}\right).$$

Observe that both coordinates of P_t are rational. Since t can take infinitely many rational values, we can now enumerate infinitely many rational points on Γ. This, in turn, allows us to generate infinitely many PPTs. For example:

- $t = 1/4$ yields the rational point $(15/17, 8/17)$ and the PPT $(15, 8, 17)$.
- $t = 1/2$ yields the rational point $(3/5, 4/5)$ and the PPT $(3, 4, 5)$.

No finite value of t will give us the point $(-1, 0)$; we would have to let t become infinite. Leaving this point out, every other rational point on Γ can be found by our scheme. For if $P \neq A$ is any rational point on Γ other than $(-1, 0)$, the slope of AP is certainly *some* rational number; and that is all we need!

Now write $t = v/u$, where u, v are coprime integers. Doing the necessary substitutions, we obtain

$$P_{v/u} = \left(\frac{u^2 - v^2}{u^2 + v^2}, \frac{2uv}{u^2 + v^2}\right),$$

and we have at hand a PT, namely $(u^2 - v^2, 2uv, u^2 + v^2)$. To ensure that this is a PPT, we only need u and v to be coprime and of opposite parity. (The verification of this is left to the reader.)

10.5 Of wise men and camels

The following story is well known. A man dictates that after his death his 17 camels should be distributed amongst his three sons in the following manner: 1/2 should go to the eldest, 1/3 to the middle one, and 1/9 to the youngest one. The sons are not too happy at the prospect of chopping up the camels, so they seek the help of a wise man. The wise man brings a camel of his own and adds it to the herd, which now has 18 camels. He then gives 1/2 of 18 or 9 camels to the eldest, 1/3 of 18 or 6 camels to the middle one, and 1/9 of 18 or 2 camels to the youngest one. The number of camels given away is now $9 + 6 + 2 = 17$. One camel is left—the one brought by the wise man. The camel and wise man return home, and the brothers live happily ever after.

This problem clearly offers scope for further study. The numbers used above were 17, 2, 3 and 9. It is clear that the method used by the wise man worked because

$$\frac{1}{2}+\frac{1}{3}+\frac{1}{9}=\frac{17}{18}, \quad \therefore \frac{1}{2}+\frac{1}{3}+\frac{1}{9}+\frac{1}{18}=1.$$

We ask the following question: *For what other combinations of positive integers is such a story possible?*

To simplify matters, let us assume that the man has only three sons. The condition he dictates is that on his death $1/a$ of the herd goes to his eldest son, $1/b$ to the middle one, and $1/c$ to the youngest; here a, b, c are positive integers and $a < b < c$. The problem acquires interest when

$$\frac{1}{a}+\frac{1}{b}+\frac{1}{c}=\frac{N-1}{N},$$

for some positive integer N. Then if the number of camels is N, the division can be done as described earlier. (Add a camel to the herd, then give away in succession $1/a$, $1/b$ and $1/c$ of the original herd to the sons. The division involves no chopping up of the camels and can be done with no pain to the camel or the owner. At the end, 1 camel is left, which then goes back to its original home. The task has been accomplished.)

The equation displayed above may be written as:

$$\frac{1}{a}+\frac{1}{b}+\frac{1}{c}+\frac{1}{N}=1.$$

We consider the following problem: *Find all quadruples* (a, b, c, d) *of positive integers that satisfy this equation, plus the condition* $a < b < c < N$.

Since $a < b < c < N$, we have $1/a > 1/b > 1/c > 1/N$, and so

$$\frac{4}{a} > \frac{1}{a}+\frac{1}{b}+\frac{1}{c}+\frac{1}{N} > \frac{4}{N}.$$

Therefore, $4/a > 1 > 4/N$, which shows that $a < 4 < N$. We also have $a > 1$; therefore, $1 < a < 4$, which means that $a = 2$ or 3. We shall pursue both possibilities.

If $a = 3$, then $1/b + 1/c + 1/N = 2/3$. Using the inequalities $b < c < N$ as we did earlier, we get $3/b > 2/3 > 3/N$, or $b < 9/2$; that is, $b \leq 4$ since b is an integer. Since $b > a$, we can only have $b = 4$. The equation now becomes $1/c + 1/N = 5/12$. The inequalities yield: $2/c > 5/12 > 2/N$, so $c < 24/5$; that is, $c \leq 4$ as c is an integer. Since we must have $c > 4$ (because $b = 4$), no solution is possible. That is, no solution is possible with $a = 3$.

If $a = 2$, then $1/b + 1/c + 1/N = 1/2$. Since $b < c < N$ we get $3/b > 1/2 > 3/N$, therefore $b < 6$. So $2 < b < 6$, which means that $b = 3$, 4 or 5. Each of these possibilities has to be considered.

If $b = 3$, then we get $1/c + 1/N = 1/6$, which yields $2/c > 1/6 > 2/N$, therefore $c < 12$. Also $1/c < 1/6$, which yields $c > 6$. So $6 < c < 12$, which means that $c = 7$, 8, 9, 10 or 11. We quickly find, by direct substitution, that $c = 11$ and $c = 12$ do not work, whereas the other possibilities yield solutions. (The equation does work with $c = 12$, but we get $N = 12$ too, which is not permitted; we must have $N > c$.) So we obtain four possible quadruples simultaneously:

$$(2, 3, 7, 42), \quad (2, 3, 8, 24), \quad (2, 3, 9, 18), \quad (2, 3, 10, 15).$$

(The third quadruple is the one that occurs in the original problem.)

If $b = 4$, then we get $1/c + 1/N = 1/4$, which yields $2/c > 1/4 > 2/N$, and therefore $c < 8$. Also $1/c < 1/4$, so we have $4 < c < 8$, i.e., $c = 5$, 6 or 7. Only $c = 7$ fails to work, and we obtain two more quadruples:

$$(2, 4, 5, 20), \quad (2, 4, 6, 12).$$

If $b = 5$, then we get $1/c + 1/N = 3/10$, which yields $2/c > 3/10 > 2/N$, and therefore $c < 20/3$. That is, $c \leq 6$ since c is an integer. Since $b = 5$ and $c > b$ we can only have $c = 6$. However, this does not yield an integral value for N (we get $N = 7\frac{1}{2}$).

So the quadruples (a, b, c, N) of positive integers with $a < b < c < N$ and $1/a + 1/b + 1/c + 1/N = 1$ are

$$\begin{array}{lll} (2, 3, 7, 42), & (2, 3, 8, 24), & (2, 3, 9, 18), \\ (2, 3, 10, 15), & (2, 4, 5, 20), & (2, 4, 6, 12). \end{array}$$

If we replace $a < b < c < N$ by $a \leq b \leq c \leq N$, then many more solutions are possible. The reader is requested to compile them.

★ ★ ★

The corresponding problem with only two sons is much easier to solve. Here we look for triples (a, b, N) of positive integers, with $a < b < N$ and $1/a + 1/b + 1/N = 1$. It turns out that there is just one solution: $(a, b, N) = (2, 3, 6)$.

The proof is simple. Since $a < b < N$ we have $3/a > 1$, $\therefore a < 3$, so $a = 2$. This yields $1/b + 1/N = 1/2$, $\therefore 2/b > 1/2 > 2/N$, giving $b < 4$ and so $b = 3$. This, in turn, yields $N = 6$.

⋆ ⋆ ⋆

The corresponding problem with *four* sons is very much more tedious to solve, and we shall not do so here.

10.6 Harder examples

In this section, we present a few number theoretic problems of a somewhat harder nature, and indicate how they may be tackled. The problems are given first, and later the solutions. The reader may wish to try the problems before proceeding to the solutions.

(1) Show that there is no positive integer n such that the quantity $8^n + 2^n + 1$ is a square.

(2) Find all squares which may be written as the sum of a power of 2 and a power of 3. Equivalently, find all pairs, (m, n), of positive integers such that the quantity $2^m + 3^n$ is a square.

(3) (This was Problem 2 of the 32nd IMO, held in Sweden in 1991.)

Let $n > 6$ be an integer, and let $a_1, a_2, \ldots, a_k$ be all the natural numbers less than n and relatively prime to n. Suppose that

$$a_2 - a_1 = a_3 - a_2 = \cdots = a_k - a_{k-1}.$$

Show that n is either a prime number or a power of 2.

(4) (This was Problem 4 of the 35th IMO, held in Hong Kong in 1994.)

Find all pairs (m, n) of positive integers greater than 1 such that $n^3 + 1$ is divisible by $mn - 1$.

⋆ ⋆ ⋆

Solution to problem 1 Suppose that $8^n + 2^n + 1 = a^2$, where a is an integer. We first read the equation relative to the modulus 3 (reading an equation modulo some convenient number is part of the standard approach in the study of such equations), and then we use the obvious fact that there cannot be a square between two consecutive squares.

Since modulo 3 we have $8 \equiv 2 \equiv -1$, we get

$$(-1)^n + (-1)^n + 1 \equiv a^2 \pmod{3}.$$

If n is odd, then this becomes $a^2 \equiv -1 \pmod{3}$, which is not possible. So n must be even. Let $n = 2m$, where m is an integer. The equation now reads: $64^m + 4^m + 1 = a^2$. Observing that

$$(8^m)^2 = 64^m, \quad (8^m + 1)^2 = 64^m + 2 \cdot 8^m + 1 > 64^m + 4^m + 1,$$

we obtain

$$(8^m)^2 < a^2 < (8^m + 1)^2;$$

equivalently, $8^m < a < 8^m + 1$, which clearly is absurd, as 8^m and $8^m + 1$ are consecutive integers. It follows that $8^n + 2^n + 1$ cannot be a square.

Solution to problem 2 Let $2^m + 3^n = a^2$, where m, n and a are positive integers.

We first read the equation modulo 3. Since 3 divides 3^n but not 2^m, it follows that 3 is not a divisor of a^2. This means that $a^2 \equiv 1 \pmod{3}$, therefore that $2^m \equiv 1 \pmod{3}$, which holds iff m is even.

Next, we read the equation modulo 4. Since m is even, we have $m \geq 2$, so $4 \mid 2^m$. Also, 4 is not a divisor of 3^n, so 4 is not a divisor of a^2 either. Therefore, $a^2 \equiv 1 \pmod{4}$, and so $3^n \equiv 1 \pmod{4}$. This means that n is even.

So both m and n are even integers, implying that 2^m and 3^n are squares. Since 2^m and 3^n are coprime, we see that $\left(2^{m/2}, 3^{n/2}, a\right)$ is a PPT. The theorem proved earlier about PPTs now applies, and we deduce that there exist positive integers u, v of opposite parity, with $u > v$, such that

$$2^{m/2} = 2uv, \quad 3^{n/2} = u^2 - v^2, \quad a = u^2 + v^2.$$

The first relation shows that u and v are both powers of 2 (the factors of a power of 2 can only be powers of 2). Since u and v

are coprime and $u > v$, this implies that $v = 1$. The relation in the middle may now be written as

$$3^{n/2} = (u-1)(u+1),$$

which shows that $u-1$ and $u+1$ are both powers of 3 (the factors of a power of 3 can only be powers of 3). The only two powers of 3 which differ by 2 are 1 and 3, so $u-1=1$ and $u+1=3$; i.e., $u=2$. This yields:

$$2^{m/2} = 4, \quad \therefore \quad 2^m = 16; \quad 3^{n/2} = 3, \quad \therefore \quad 3^n = 9; \quad a = 5.$$

So there is precisely one square which may be written as a sum of a power of 2 and a power of 3, namely 5^2:

$$5^2 = 25 = 16 + 9 = 2^4 + 3^2.$$

Solution to problem 3 Clearly, $a_1 = 1$ and $a_k = n-1$ (for 1 and $n-1$ are coprime to n). Let $d = a_2 - a_1$. Then

$$a_2 = 1 + d, \quad a_3 = 1 + 2d, \quad \ldots, \quad a_k = 1 + (k-1)d,$$

so $n - 1 = 1 + (k-1)d$, i.e., $n = 2 + (k-1)d$.

Suppose that $d = 1$. Then

$$a_1 = 1, \quad a_2 = 2, \quad a_3 = 3, \quad \ldots, \quad a_{k-1} = n - 1,$$

so *all* positive integers less than n are coprime with n. This clearly implies that n is prime.

Suppose that $d = 2$. Then

$$a_1 = 1, \quad a_2 = 3, \quad a_3 = 5, \quad \ldots,$$

so all odd positive integers less than n are coprime with n. This means that n possesses no odd prime factor, implying that n is a power of 2.

Now suppose that $d > 2$. Then $a_2 \geq 4$, so 3 is not coprime with n; $\therefore\ 3 \mid n$. Since $n = 2 + (k-1)d$, we obtain

$$2 + (k-1)d \equiv 0 \pmod 3, \qquad \therefore \quad (k-1)d \equiv -1 \pmod 3.$$

So $(k-1)d$ is not a multiple of 3, which implies in particular that 3 is not a divisor of d. So $d \equiv 1$ or $2 \pmod 3$.

If $d \equiv 2 \pmod 3$ then $2 \mid 1+d$, and if $d \equiv 1 \pmod 3$ then $3 \mid 1+2d$. So in any case, $3 \mid a_2$ or a_3. But this cannot happen: 3 is a divisor of n, so none of the as can be a multiple of 3.

We conclude that d cannot be greater than 2. There are only two possibilities left, both of which we have considered. The desired conclusion follows.

Solution to problem 4 The condition is $mn-1 \mid n^3+1$. Since

$$n^3(m^3+1)-(n^3+1)=n^3m^3-1=(mn)^3-1,$$

and $mn-1 \mid (mn)^3-1$, we deduce that

$$mn-1 \mid n^3(m^3+1)-(n^3+1).$$

So if $mn-1 \mid n^3+1$, then $mn-1 \mid n^3(m^3+1)$. But $mn-1$ and n^3 are coprime; for if a prime p is a divisor of n^3, then we have $p \mid n$, $\therefore p \mid mn$, $\therefore p$ is not a divisor of $mn-1$. Therefore, from $mn-1 \mid n^3(m^3+1)$, we can conclude that $mn-1 \mid m^3+1$. So if $mn-1$ is a divisor of n^3+1 then it is also a divisor of m^3+1. Thus, a symmetry exists between m and n which is not obvious in the statement of the problem. So we may suppose, without any loss, that $m \geq n$.

We first consider the trivial cases. If $m=n$ then the condition becomes $n^2-1 \mid n^3+1$, or $n-1 \mid n^2-n+1$, or $n-1 \mid 1$ (since $n-1 \mid n^2-n$). This yields $n-1=1$ and, therefore, $n=2$. We obtain the solution $(m,n)=(2,2)$.

Now we take $m>n$. If $n=1$ then the condition becomes $m-1 \mid 2$, which yields $m-1=1$ or 2, i.e., $m=2$ or 3. So we obtain two more solutions: $(m,n)=(2,1)$ and $(3,1)$.

Now assume that $m>n>1$. Let $k=(n^3+1)/(mn-1)$. Then k is an integer, and

$$n^3+1=kmn-k, \quad \therefore \ 1 \equiv -k \pmod n.$$

Since $k \equiv -1 \pmod n$, we may write $k=an-1$ for some positive integer a. We now have

$$\frac{n^3+1}{mn-1}=an-1.$$

Now $mn-1>n^2-1$. Replacing the denominator $mn-1$ by the smaller quantity n^2-1, we see that

$$an-1<\frac{n^3+1}{n^2-1}=n+\frac{1}{n-1}.$$

Since $an-1$ is an integer and $n>1$ this yields $an-1 \le n$, which implies that $a=1$. Therefore,

$$\frac{n^3+1}{mn-1} = n-1, \quad \therefore \quad (mn-1)(n-1) = n^3+1,$$

and we get, on solving for m,

$$m = \frac{n^2+1}{n-1} = n+1+\frac{2}{n-1}.$$

Thus, $n-1 \mid 2$, which yields $n-1 = 1$ or 2, and so $n=2$ or 3. Both possibilities yield $m=5$. So we get two more solutions: $(m,n) = (5,2)$ and $(5,3)$.

Recalling that the roles of m and n can be switched (we had assumed at the start that $m \ge n$), we see that the given equation has the following nine solutions: $(m,n) = (2,2)$, $(1,2)$, $(2,1)$, $(1,3)$, $(3,1)$, $(2,5)$, $(5,2)$, $(3,5)$ and $(5,3)$.

10.7 Cyclic numbers

Consider the number 142857, which is the repeating portion of the base-10 decimal expansion of the fraction 1/7. If we multiply it by 2, 3, 4, 5 and 6 respectively, we get the numbers 285714, 428571, 571428, 714285 and 857142. Each of these is, digit-wise, a cyclic permutation of the original number; a curious property indeed! Significantly, if we multiply the number by 7 we get 999999. We shall study such numbers in this section and look for ways of finding them.

Let the positive integer N have the base-10 representation

$$N = \overline{\mathsf{a}_1\,\mathsf{a}_2\,\mathsf{a}_3\,\ldots\,\mathsf{a}_{n-1}\,\mathsf{a}_n} = \sum_{i=1}^{n} \mathsf{a}_i\, 10^{n-i},$$

where the a_i are digits, and let N' be the integer obtained by bringing the units digit of N to the front; i.e.,

$$N' = \overline{\mathsf{a}_n\,\mathsf{a}_1\,\mathsf{a}_2\,\mathsf{a}_3\,\ldots\,\mathsf{a}_{n-1}}.$$

Suppose that we are told that $N' = 2N$. Can we find N using only this information? Yes! Here is an elegant approach which yields *all* such Ns (there are infinitely many of them).

We consider the infinite recurring decimals x and y whose repeating portions are the numbers N and N' respectively:

$$\begin{aligned} x &= 0.\overline{a_1a_2\ldots a_{n-1}a_n}\ \overline{a_1a_2\ldots a_{n-1}a_n}\ \overline{a_1\ldots}\ \ldots, \\ y &= 0.\overline{a_na_1a_2\ldots a_{n-1}}\ \overline{a_na_1a_2\ldots a_{n-1}}\ \overline{a_n\ldots}\ \ldots. \end{aligned}$$

Since $N' = 2N$, it follows that $y = 2x$. If we multiply y by 10, we get

$$10y = a_n.\overline{a_1a_2\ldots a_{n-1}a_n}\ \overline{a_1a_2\ldots a_{n-1}a_n}\ \overline{a_1\ldots}\ \ldots = a_n + x,$$

implying that $20x = a_n + x$ and $x = a_n/19$. Now a_n can take any value between 2 and 9 (it cannot be 1). Each choice yields an answer! For instance, $a_n = 2$ yields $x = 2/19$. Now, by actual division, we get

$$\frac{2}{19} = 0.\overline{105263157894736842}\ \overline{105263157894736842}\ \ldots.$$

The integer N is clearly the repeating portion of x, so we get

$$N = 105263157894736842.$$

The reader should check that N does indeed have the required property. Note that this is the *least* number with this property. This is clear, since a_n cannot be less than 2.

Interestingly, the choices $a_n = 3, 4, \ldots, 8, 9$ yield numbers which are all cyclic permutations of the number just obtained; e.g., $a_n = 7$ yields:

$$N = 368421052631578947$$

The cyclic property just mentioned may be checked by the reader.

★ ★ ★

If we require that $N' = 3N$ then, with x and y defined as above, we obtain the equation $30x = a_n + x$, or $x = a_n/29$. Once again, each choice for a_n (but with $a_n > 2$) yields a corresponding N. Thus, $a_n = 3$ yields:

$$\frac{3}{29} = 0.\overline{1034482758620689655172413793}\ \ldots,$$

and we obtain

$$N = 1034482758620689655172413793.$$

If we take $a_n = 8$, we get

$$N = 2758620689655172413793103448.$$

As earlier, different choices for a_n yield numbers which are, digit-wise, cyclic permutations of one another.

⋆ ⋆ ⋆

The analysis may easily be extended to the case when we transfer two digits instead of one. For instance, if we seek an integer N such that $N' = 2N$, where N' is obtained from N by shifting the last two digits to the front (but preserving the cyclic order of the digits), then we write $N = \overline{a_1\, a_2\, a_3\, \ldots\, a_{n-1}\, a_n}$ as earlier, and

$$\begin{aligned} x &= 0.\overline{a_1 a_2 \ldots a_{n-2} a_{n-1} a_n}\ \overline{a_1 a_2 \ldots a_{n-2} a_{n-1} a_n}\ \overline{a_1 \ldots}\ \ \ldots, \\ y &= 0.\overline{a_{n-1} a_n a_1 a_2 \ldots a_{n-2}}\ \overline{a_{n-1} a_n a_1 a_2 \ldots a_{n-2}}\ \overline{a_{n-1} \ldots}\ \ \ldots. \end{aligned}$$

We now get $y = 2x$ and also $100y = \overline{a_{n-1} a_n} + x$. Let A be the 2-digit base-10 number $\overline{a_{n-1} a_n}$. Then $200x = A + x$, $\therefore\ x = A/199$. Different choices for A yields corresponding values for N. For example, $A = 25$ yields a value of N with 99 digits (whew!):

$$\begin{aligned} N \;=\; & 125628140703517587939698492462311557788944723618090 \\ & 452261306532663316582914572864321608040201005025. \end{aligned}$$

Multiplication by 2 yields the number 25125628 ... 010050 (the '25' has moved to the 'front').

10.8 Binomial coefficients

In this section, we shall study some of the arithmetical properties of the binomial coefficients. We ask the following question:

> *Given positive integers n and i, with $i \le n$, how do we ascertain whether $\binom{n}{i}$ is odd or even; that is, how do we ascertain the parity of this number?*

Interestingly, there are two distinct algorithms available, but (naturally) they are equivalent to one another. Each has it own appeal.

A recursive algorithm In the first, we assume that $i \le n/2$ (if not, we first replace i by $n - i$). The algorithm is now implied by the following extremely interesting result. *Let 2^k be the largest power of 2 below n (if n is a power of 2 then $2^k = n$). Then the two numbers*

$$\binom{n}{i}, \quad \binom{n-2^k}{i},$$

have the same parity.

The result immediately gives rise to a recursive algorithm. For instance, say we wish to know the parity of $\binom{100}{20}$. Since 64 is the largest power of 2 below 100, and $100 - 64 = 36$, we infer that the numbers

$$\binom{100}{20}, \quad \binom{36}{20}$$

have equal parity. Next, $\binom{36}{20} = \binom{36}{16}$, and the largest power of 2 below 36 is 32. Since $36 - 32 = 4$ we infer that $\binom{36}{16}$ and $\binom{4}{16}$ have the same parity. But $\binom{4}{16} = 0$, an even number; it follows that $\binom{100}{20}$ is even. (Skeptics who doubt the result should call upon the ever-reliable *Mathematica*; it reveals that

$$\binom{100}{20} = 535983370403809682970,$$

which is clearly even.)

Similarly, in checking the parity of $\binom{200}{50}$, we obtain the following coefficients *en route* to the final answer:

$$\binom{200}{50}, \quad \binom{72}{50}, \quad \binom{72}{22}, \quad \binom{8}{22}.$$

The last quantity is 0, implying that $\binom{200}{50}$ is even.

⋆ ⋆ ⋆

An immediate corollary of the result is the following: *If k is any positive integer, then the numbers*

$$\binom{2^k}{1}, \quad \binom{2^k}{2}, \quad \ldots, \quad \binom{2^k}{2^k-2}, \quad \binom{2^k}{2^k-1}$$

are all even.

An algorithm based on base-2 representations The second algorithm is suggested by another very interesting and pretty result. *The number $\binom{n}{i}$ is odd iff in the binary representations of i and $n-i$ the locations of the 1s have no overlap; equivalently, iff in computing the binary sum of i and $n-i$ there is no 'carry'.*

For example, to check the parity of $\binom{100}{20}$, we first compute the binary forms of 20 and 80; we get

$$\begin{aligned} 80 &= (1010000)_2, \\ 20 &= (0010100)_2. \end{aligned}$$

Observe that there *is* an overlap—in the third place (both numbers have a 1). We deduce that $\binom{100}{20}$ is even.

In checking the parity of $\binom{15}{5}$, we get

$$\begin{aligned} 10 &= (1010)_2, \\ 5 &= (0101)_2. \end{aligned}$$

No overlap occurs this time; the addition can be done without any 'carry'. It follows that $\binom{15}{5}$ is odd. (The number is easily computed to be 3003, which is indeed odd.)

⋆ ⋆ ⋆

Remarkably, the second result can be greatly generalized; indeed we have the following very strong statement due to Ernst Kummer (1810–1893; best-known to number theorists for his outstanding work on 'Fermat's last theorem'):

> *Given a prime number p and positive integers n and i, with $i \le n$, the number $\binom{n}{i}$ is indivisible by p iff in computing the base-p sum of i and $n-i$ no 'carry' is met at any stage.*

For example, to check the divisibility of $\binom{100}{20}$ by 3 we first compute the base-3 forms of 80 and 20:

$$\begin{aligned} 80 &= (2222)_3, \\ 20 &= (0202)_3. \end{aligned}$$

A 'carry' is quite visibly encountered when we add $(2222)_3$ and $(0202)_3$; we infer that $\binom{100}{20}$ is divisible by 3.

On the other hand, consider the number $\binom{8}{2}$. The base-3 forms of 6 and 2 are $(20)_3$ and $(02)_3$ respectively, and the two numbers can clearly be added with no 'carry'. This implies that $\binom{8}{2}$ is not divisible by 3. (Indeed, $\binom{8}{2} = 28$, which is indivisible by 3.)

We leave the proofs to the problem set in the next section.

10.9 Problems

10.9.1 Prove that $\sqrt{3}$, $\sqrt{5}$ and $\sqrt{7}$ are irrational.

10.9.2 Prove, more generally, that $\sqrt{p}$ is irrational for all prime numbers p.

10.9.3 Prove that $\sqrt[3]{2}$ is irrational.

10.9.4 Prove, more generally, that for any two integers m and N, if $\sqrt[m]{N}$ is not an integer then it is irrational. In other words, $\sqrt[m]{N}$ is never a non-integral rational number.

10.9.5 Prove that $\sqrt{2}+\sqrt{3}$ is irrational.

10.9.6 Prove that the following numbers are irrational:
(a) $\log_{10} 2$ (b) $\log_{10} 3$.

10.9.7 Prove that for positive integers a and b exceeding 1, the quantity $\log_a b$ is rational if and only if both a and b are integral powers of a common positive integer c.

(**EXAMPLE.** $\log_9 27$ is rational, but not $\log_9 16$.)

10.9.8 Let a, b, c be odd integers. Prove that the roots of the equation $ax^2+bx+c=0$ are irrational.

10.9.9 Let $r \approx 1.25992$ represent the cube root of 2. Find all triples (a, b, c) of integers for which $ar^2+br+c=0$.

10.9.10 Find all triples (a, b, c) of integers that satisfy the equation $a+b\sqrt{2}+b\sqrt{3}=0$.

10.9.11 Prove that for any integer n:

(a) $n^2 \equiv 0$ or $1 \pmod 3$;
(b) $n^2 \equiv 0$ or $1 \pmod 4$;
(c) $n^2 \equiv 0$, 1 or $4 \pmod 5$;
(d) $n^2 \equiv 0$, 1, 2 or $4 \pmod 7$.

10.9.12 Show that if n is an odd integer then $n^2 \equiv 1 \pmod 8$.

10.9.13 Do rational numbers u, v exist for which $u^2+v^2=3$?

10.9.14 (Continuation) More generally, we ask the following:

Let u and v be rational numbers. What integral values can $z = u^2 + v^2$ take?

Certain integral values are obviously achievable, e.g.:

- $z = 1$, from $u = 3/5$ and $v = 4/5$;
- $z = 2$, from $u = v = 1$;
- $z = 4$, from $u = 2$ and $v = 0$;
- $z = 5$, from $u = 2$ and $v = 1$;

and so on. However, it appears that z never takes certain values: 3, 6, 7, ... How can this be shown?

10.9.15 (Continuation) Use the above result to find another proof of the assertion that $\sqrt{2}$ is irrational.

HINT. Suppose that $a^2 = 2b^2$ for two non-zero integers a and b. Adding b^2 to both sides yields the relation $a^2 + b^2 = 3b^2$. What can be concluded from this?

10.9.16 (Continuation) Use the theme of the above proof to show that the following quantities too are irrational: $\sqrt{3}$, $\sqrt{5}$ and $\sqrt{7}$.

10.9.17 In this problem, we encounter yet another proof of the irrational nature of $\sqrt{2}$, based on the principle of *infinite descent*—a technique pioneered by Fermat. The underlying idea is simply that it is not possible to have an infinitely long and strictly decreasing sequence of positive integers. (Sooner or later the sequence will reach 1, and then it can proceed no further.)

Suppose that $\sqrt{2} = \dfrac{a}{b}$, where a, b are positive integers.

(a) Show that $b < a < 2b$, hence that $0 < 2b - a < a$ and $0 < a - b < b$.

(b) Show that from $\sqrt{2} = a/b$ there follows

$$\sqrt{2} = \frac{2b - a}{a - b}.$$

(c) Deduce that we have found a fraction that equals $\sqrt{2}$ but has a smaller numerator as well as a smaller denominator than a/b.

How does this line of reasoning establish the irrational nature of $\sqrt{2}$?

10.9.18 Find all pairs (x, y) of positive integers for which $x^2 - y^2 = 1000$.

10.9.19 (Continuation) Show that there are infinitely many positive integers n for which the equation $x^2 - y^2 = n$ has no solutions in integers x, y.

10.9.20 (Continuation) Suppose that the equation $x^2 - y^2 = n$ does possess solutions in positive integers. Let f_n denote the number of solutions ($f_3 = f_5 = 1$, $f_{60} = 2$, because $60 = 8^2 - 2^2 = 16^2 - 14^2$). Find a way of computing f_n for some arbitrary n.

10.9.21 Let d_n denote the number of positive divisors of the integer n (we include 1 and n in the count.) Thus, $d_{10} = 4$ (the divisors being 1, 2, 5 and 10), $d_{12} = 6$ (the divisors being 1, 2, 3, 4, 6 and 12), and so on.

Classify the integers n for which: (a) $d_n = 2$; (b) $d_n = 3$; (c) $d_n = 4$; (d) $d_n = 5$.

10.9.22 Find all positive integers $n < 10^{1000}$ for which $d_n = 1993$.

10.9.23 Determine the units digits of each of the following:
(a) 2^{100} (b) 2^{1995} (c) 7^{1995} (d) 1993^{1994}.

10.9.24 Determine the last two digits of 2^{100}.

10.9.25 Determine the number of zeroes at the units end of 100!. (By definition, $N! = 1 \times 2 \times 3 \times \cdots \times N$.)

10.9.26 Determine the least positive integer n for which

(a) $n!$ is divisible by 1995;

(b) $n!$ is divisible by 1999;

(c) $n!$ has precisely 1993 zeroes at its units end.

10.9.27 Let (a, b, c) be a PPT. Show that: (a) $2 \mid ab$; (b) $3 \mid ab$; (c) $4 \mid ab$; (d) $5 \mid abc$; (e) $60 \mid abc$.

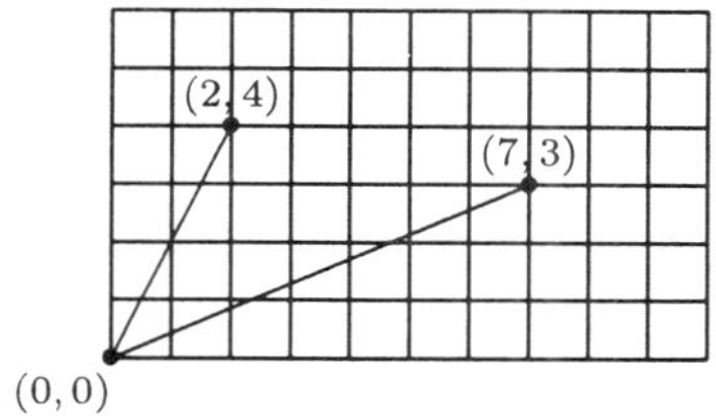

Figure 10.2. *Grid crossing*

10.9.28 Observe that for any integer n, if n has 6 for its units digit, then so does n^2; if n has 25 for its last two digits, then so does n^2. Numbers such as 6 and 25 are referred to as *automorphic numbers*. So if n terminates in the string of digits corresponding to an automorphic number, then n^2 terminates in the same string.

(a) Determine all 1-digit automorphic numbers.
(b) Determine all 2-digit automorphic numbers.
(c) Determine all 3-digit automorphic numbers.

10.9.29 Does there exist an integer A for which A and A^2 contain, when considered together, each of the 9 non-zero digits, once each?

10.9.30 Show that $2^n + 1$ is never a multiple of 7.

10.9.31 Show that if $7^n + 1$ is a multiple of 13 then n is an odd multiple of 6.

10.9.32 Find all pairs (m, n) of integers for which $2^m - 3^n = \pm 1$.

10.9.33 **Crossing squares.** In this problem, we consider the grid created on the coordinate plane by the lines $x = 0$, $x = 1$, $\ldots$, $y = 0$, $y = 1$, $\ldots$; the lines divide the plane into unit squares or cells (see Figure 10.2).

Let m, n be positive integers, and let a line segment be drawn joining the points $(0, 0)$ and (m, n). Let $C_{m,n}$ denote the number of cells through which the segment crosses. (The segment should pass through the *interior* of a cell for that cell to be included in the count.)

EXAMPLE. $C_{2,4} = 4$ (see Figure 10.2), and $C_{7,3} = 9$.

Find a formula for $C_{m,n}$ in terms of m, n.

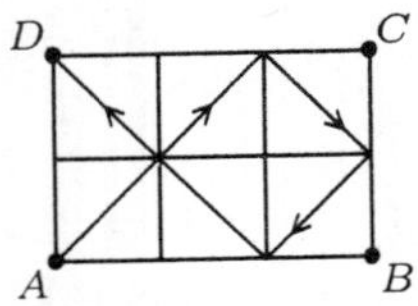

a 3 × 2 table

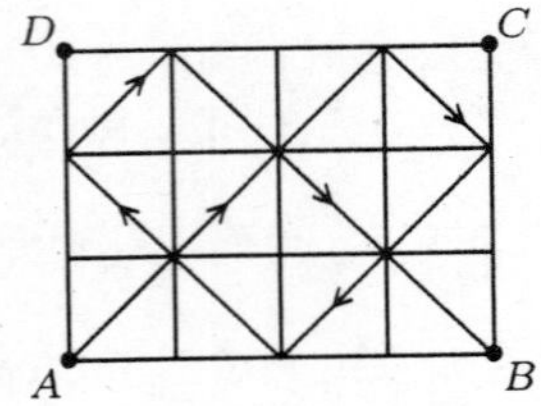

a 4 × 3 table

Figure 10.3. *Digital billiards*

10.9.34 Digital billiards. Consider an $m \times n$ rectangular billiards-table (m, n are positive integers) with a hole at each corner. The corners are labelled A, B, C, D as shown in Figure 10.3. A billiard ball is placed at A and struck in the north-easterly direction (i.e., at an angle of 45° to the side of the table).

NOTE. We assume that the table is smooth ('motion with no friction') and that reflections take place off the walls as dictated by the laws of reflection of light ('angle of incidence equals angle of reflection').

The figures above show what happens on a 3×2 board (after 3 reflections the ball reaches the D-hole), and on a 4×3 board (after 5 reflections the ball reaches the B-hole). We therefore write:

$$R_{3,2} = 3, \quad D_{3,2} = D$$
$$R_{4,3} = 5, \quad D_{4,3} = B.$$

(R for 'number of reflections', D for 'destination').

Find formulas for $R_{m,n}$ and $D_{m,n}$ in terms of m and n.

10.9.35 Prove that if a, b, c, d and a', b', c', d' are positive integers such that

$$a + \cfrac{1}{b + \cfrac{1}{c + \cfrac{1}{d}}} = a' + \cfrac{1}{b' + \cfrac{1}{c' + \cfrac{1}{d'}}},$$

then $a = a'$, $b = b'$, $c = c'$ and $d = d'$.

COMMENT. Descending fractions of the type shown above are referred to as *simple continued fractions*.

10.9.36 (This is a problem that most people have come across in some form or other.)

We have before us a large tank of wine and two vessels, of which one can hold 3 liters of liquid while the other can hold 5 liters. Using only these vessels, how can one take away exactly 1 liter of wine?

Note that *estimation is disallowed*—we are allowed to work only with vessel-fulls at a time. We must find a sequence of exchanges, pouring from one vessel into the other and so on, at the end of which the desired 1 liter of wine in left in one of the vessels. (Assume that the tank contains an unlimited amount of wine.)

Solve the problem. Invent your own notation and write down your solution using the notation.

10.9.37 (Continuation) Generalize the problem. Let the two vessels have capacities a liters and b liters, respectively ($0 < a < b$, and a, b are integers). Are there combinations of a and b for which the end-result (exactly 1 liter to be left in one vessel) can never be attained?

10.9.38 (Continuation) Assuming that the end-result is attainable, show how we can reach it. Look for an algorithm that works for all a, b (provided, of course, that a solution is possible).

10.9.39 (Continuation) For given values of a, b, let $n_{a,b}$ denote the least number of steps needed to measure out 1 liter, with each act of filling, refilling or emptying counting as one step.

EXAMPLE. $n_{3,2} = 3$ (fill the '3', empty it into the '2', empty the '2' back into the tank; 1 liter remains in the '3').

If, at an intermediate stage, we have x liters of wine in the 'a'-jug and y liters in the 'b'-jug, we describe the state as (x, y). Starting from $(0, 0)$ (both vessels empty), we must reach $(1, 0)$ or $(0, 1)$. If for some a, b there is no way of obtaining 1 liter, we write $n_{a,b} = \infty$.

Find a formula for $n_{a,b}$ in terms of a, b.

10.9.40 Find ways of factorizing moderately large numbers (having four to six digits), and use your approach to factorize the following numbers: (a) 28891 (b) 99899 (c) 403189.

HINT. Use the identity $a^2 - b^2 = (a+b)(a-b)$.

10.9.41 Find all pairs (x, y) of positive integers (or show that none exist) such that:
(a) $\frac{1}{x} + \frac{1}{y} = \frac{19}{96}$ (b) $\frac{1}{x} + \frac{1}{y} = \frac{19}{97}$.

10.9.42 Let n be an arbitrary positive integer. Show that if $2^n - 1$ is prime then n is prime.

10.9.43 Let n be an arbitrary positive integer. Show that if $2^n + 1$ is prime then n is a power of 2.

10.9.44 Find all n for which $2^n - 1$ and $2^n + 1$ are both prime.

10.9.45 $\boxed{*}$ Let n be an odd positive integer. Is it true that if $(2^n + 1)/3$ is prime, then so is n?

10.9.46 Show that $4^n + 2$ is a multiple of 6 for all positive integers n.

10.9.47 Find all positive integers n for which $n^4 + 4$ is prime.

10.9.48 Show that for $n > 1$ the number $4^n + n^4$ is never prime.

10.9.49 Does there exist an infinite subset A of positive integers such that:

(a) no subset of A adds up to a square?
(b) no subset of A adds up to a cube?

10.9.50 For any positive integer n, let $(1 + \sqrt{2})^n$ be simplified and written in the form $x + y\sqrt{2}$, where x, y are positive integers. Show that $x^2 - 2y^2 = \pm 1$.

EXAMPLE. When $n = 2$ the expression simplifies to $3 + 2\sqrt{2}$, so $x = 3, y = 2$. Observe that $3^2 - 2(2^2) = 1$.

10.9.51 (Continuation) Let x and y be positive integers such that $x^2 - 2y^2 = \pm 1$. Show that there exists a positive integer n such that:

$$x + y\sqrt{2} = \left(1 + \sqrt{2}\right)^n.$$

(Note that this provides us with a complete characterization of the positive integral solutions of the equation $x^2 - 2y^2 = \pm 1$.)

10.9.52 (Continuation) Find a similar characterization of the positive integral solutions of the equation $x^2 - 5y^2 = 1$.

10.9.53 For positive integers n, let $b(n)$ denote the *number of ones in the binary representation of* n.

EXAMPLE. $b(5) = 2$ because $5 = (101)_2$; $b(6) = 2$ because $6 = (110)_2$; and $b(7) = 3$ because $7 = (111)_2$.

Prove the following: for all positive integers n,

$$\left[\frac{n}{2}\right] + \left[\frac{n}{4}\right] + \left[\frac{n}{8}\right] + \left[\frac{n}{16}\right] + \cdots = n - b(n).$$

EXAMPLE. Let $n = 21$; then the sum on the left is $10 + 5 + 2 + 1 = 18$. Next, $21 = (10101)_2$, so $b(21) = 3$. Observe that $18 = 21 - 3$.

10.9.54 Can any analogous statement be made about the sum

$$\left[\frac{n}{3}\right] + \left[\frac{n}{9}\right] + \left[\frac{n}{27}\right] + \left[\frac{n}{81}\right] + \cdots,$$

for an arbitrary positive integer n?

10.9.55 Can any analogous statement be made about the sum

$$\left[\frac{n}{4}\right] + \left[\frac{n}{16}\right] + \left[\frac{n}{64}\right] + \left[\frac{n}{256}\right] + \cdots,$$

for an arbitrary positive integer n?

10.9.56 Find a generalization suggested by the results of the last three problems.

10.9.57 For positive integers n, let $f(n)$ denote the *integer closest to* n, with the convention that if the decimal part of n is 0.5 then we round *up*.

EXAMPLE. $f(2.7) = 3$, $f(3.5) = 4$, and so on. It may be checked that $f(n) = [n + 0.5]$.

Prove the following: for all positive integers n:

$$f\left(\frac{n}{2}\right) + f\left(\frac{n}{4}\right) + f\left(\frac{n}{8}\right) + f\left(\frac{n}{16}\right) + \cdots = n.$$

EXAMPLE. For $n = 21$ the sum on the left is $11+5+3+1+1 = 21$, in agreement with the claim.

10.9.58 Can any corresponding statement be made for the sum

$$f\left(\frac{n}{3}\right) + f\left(\frac{n}{9}\right) + f\left(\frac{n}{27}\right) + f\left(\frac{n}{81}\right) + \cdots ?$$

Here f refers to the same function as in the previous problem, and n is any arbitrary positive integer.

Problems concerning base-10 representations of numbers The next few problems refer to representations of numbers in the base-10 system.

10.9.59 Find the 5th digit from the right in the number

$$5^{5^{5^5}}$$

10.9.60 Show that a square cannot consist solely of odd digits.

10.9.61 (a) Is there a power of 2 whose digits can be rearranged to give another power of 2?

(b) Is there a power of 7 whose digits can be rearranged to give another power of 7?

10.9.62 Show that the tens digit of 3^n is always even.

EXAMPLE. The tens digit of 3^8 is 6, which is even ($3^8 = 6561$).

10.9.63 Show that there are no squares, cubes or fourth powers whose base-10 representations have the following forms:

$$\overline{\text{ABCABC}}, \quad \overline{\text{ABABAB}}.$$

10.9.64 Let R_n denote the number with n occurrences of '1':

$$R_n = \underbrace{111\ldots11}_{n \text{ digits}}.$$

(a) Show that if R_n is prime then so is n.

(b) Find a prime number n for which R_n is composite.

Problems concerning binomial coefficients and factorials

10.9.65 Let n and i be positive integers with $i \leq n/2$, and let 2^k be the largest power of 2 below n (if n is a power of 2, then $2^k = n$). Show that:

$$\binom{n}{i} \equiv \binom{n-2^k}{i} \pmod{2}.$$

10.9.66 Let n and i be positive integers with $i \leq n$. Show that if the 1s in the base-2 forms of i and $n - i$ occur in disjoint positions, then $\binom{n}{i}$ is odd. Conversely, show that if an overlap occurs then $\binom{n}{i}$ is even.

10.9.67 Show how the above result implies that the number $\binom{2^k}{i}$ is even for $i = 1, 2, 3, \ldots, 2^k - 1$.

10.9.68 $\boxed{*}$ Prove Kummer's result:

For positive integers n and i, with $i \leq n$, the number $\binom{n}{i}$ is indivisible by a prime p iff the base-p forms of i and $n - i$ can be added with no 'carry'.

10.9.69 Is it true that the number $\binom{3^k}{i}$ is divisible by 3 for $i = 1, 2, 3, \ldots, 3^k - 1$?

10.9.70 $\boxed{*}$ Prove the following remarkable result, due to Paul Erdös.

For integers i, j, n with $0 < i < j < n$, the numbers $\binom{n}{i}$ and $\binom{n}{j}$ possess a common factor exceeding 1.

10.9.71 Prove that for any positive integer n, the number of indices $i \in \{0, 1, 2, \ldots, n\}$ for which the quantity $\binom{n}{i}$ is odd is a power of 2.

EXAMPLE. The binomial coefficients corresponding to $n = 6$ are 1, 6, 15, 20, 15, 6, 1; and $4 = 2^2$ of these numbers are odd. If $n = 9$ then the binomial coefficients are 1, 9, 36, 84, 126, 126, 84, 36, 9, 1; and $4 = 2^2$ of these numbers are odd.

10.9.72 Prove that the exponent k of the highest power p^k of a prime p that divides the factorial number $n!$ is given by

$$k = \left[\frac{n}{p}\right] + \left[\frac{n}{p^2}\right] + \left[\frac{n}{p^3}\right] + \cdots.$$

10.9.73 Show that for all positive integers n, the quantity

$$\frac{(2n)!}{n!\,(n+1)!}$$

is an integer. (Equivalently, the binomial coefficient $\binom{2n}{n}$ is divisible by $n+1$.)

COMMENT. This problem is much simpler than it looks!

10.9.74 Let $n \geq 3$ be an integer. Show that the quantity

$$\frac{(3n)!}{n!\,(n+1)!\,(n+2)!}$$

is an integer.

10.9.75 [∗] (From the USA Mathematical Olympiad of 1975.)

Prove that the quantity

$$\frac{(5m)!\,(5n)!}{m!\,n!\,(3m+n)!\,(3n+m)!}$$

is an integer for all positive integers m and n.

10.10 References

Number theory is a vast subject, full of material for the amateur and the professional alike. In this brief survey, we have barely skimmed its surface. Here are some directions for possible study.

- The study of tests of divisibility: by 2, 4, 8, 16, ...; by 5, 25, 125, ...; by 3, 9, 27, ...; by 11; by 6, 99, ...; and by 'unfamiliar' divisors such as 7, 13, 17, 19, ...

- The study of the Fibonacci number sequence,

$$\Big\langle 1, 1, 2, 3, 5, 8, 13, 21, 34, 55, 89, 144, \ldots \Big\rangle,$$

in which each term after the second is the sum of the preceding two terms; and of other similarly defined sequences. (Such sequences are often exceedingly rich in patterns.)

- The study of indeterminate equations for which integral solutions are sought; e.g., equations such as:

$$x^2 + y^2 = z^2, \quad x^2 - 2y^2 = 1.$$

Such equations are known as *Diophantine equations*, and perhaps the best-known example is the one occurring in Fermat's Last Theorem ($x^n + y^n = z^n$ for $n > 2$).

- The study of the prime number sequence and questions related to primality: Fermat primes, Mersenne primes, perfect numbers, primes in arithmetic progressions, factorizability, verification of primality, ...

- The study of congruences: the classical theorems of Fermat and Wilson; the Chinese Remainder Theorem; quadratic residues; primitive roots; . . . .

- The study of partitions: theorems such as "the number of ways of writing a number n as a sum of *odd* numbers is equal to the number of ways of writing n as a sum of *distinct* numbers"; congruence properties of the partition function; . . . .

Further reading The reader may want to study the subject further from the following books:

[1] Andrews, G. *Number Theory.*

[2] Barnard & Child. *Higher Algebra* (Macmillan).

[3] Hardy, G H & Wright, E M. *Introduction to the Theory of Numbers* (Cambridge University Press).

[4] Niven, I & Zuckerman, H S. *Introduction to the Theory of Numbers.*

[5] Shirali, S A. *First Steps In Number Theory–1: A Primer On Divisibility* (Universities Press).

[6] Shirali, S A. *A Primer On Number Sequences* (Universities Press).

[7] Uspensky, J V. *Number Theory.*

COMMENTS. [1], [3], [4] have a lot of material on the theory of partitions. [3], [4] are highly recommended; [4] has an excellent collection of problems. [2] is excellent for algebra in general (especially algebra of the 'old-fashioned' kind) and it has a chapter on number theory as well. [5] and [6] too are highly recommended (but modesty forbids me to say more!).

Chapter 11

Coin Weighing Problems

11.1 Introduction

The class of problems referred to in the title is well-known, and it includes problems that have now become classics. As the reader will see shortly, there is a lot of pretty mathematics hidden in these problems. We start with one that is well-known.

> **The Nine-Coin Problem** *Given an accurate 2-pan balance and 9 coins of identical appearance, of which one weighs less than the others (it is counterfeit), identify the faulty coin using the balance no more than twice.*

EXERCISE

11.1.1 Solve the nine-coin problem described above. (No other item is available; the only course of action available to us is that of weighing sets of coins against one another.)

⋆ ⋆ ⋆

We proceed to investigate in detail the n-coin version of the problem ($n > 1$, as the case $n = 1$ is meaningless). Let $f(n)$ be the least number of weighings needed to find the counterfeit coin.

COMMENT. The definition of f is in terms of a best-possible method and a worst-case outcome. No element of luck is assumed, and we are to be guaranteed of a successful result after a maximum of $f(n)$ usages of the pan balance, no matter what happens during the intermediate stages.

⋆ ⋆ ⋆

Our quest is to find a formula for $f(n)$; more generally, an *algorithm* for computing $f(n)$. (Actually, 'formula' and 'algorithm' mean much the same thing.) Problem 1 suggests that $f(9) = 2$, but is this really so? Are we certain that some method does not exist whereby one weighing suffices? This is not obvious, so all we can assert at the outset is that $f(9) \leq 2$. We now proceed to show, via recursion, that indeed $f(9) = 2$.

To start with, $f(2) = f(3) = 1$, for we certainly cannot do with less than 1 weighing in either case; and on the other hand, 1 weighing suffices in both cases. (When $n = 2$, we weigh the two coins against one another; the lighter one is counterfeit. When $n = 3$, we weigh any two coins against one another; if they balance, the third coin is counterfeit, or else the lighter of the two is the faulty coin.)

It turns out to be convenient to write $f(1) = 0$. In the discussion below, we use the following notation: the kth weighing will be denoted by W_k.

Consider the case $n > 3$. Suppose that for W_1 we weigh one set of coins, A, against another set, B. We must ensure that $|A| = |B|$, otherwise the weighing will tell us nothing. Let C be the set of coins not used in the weighing. After W_1 has been done, we know with certainty which of the sets A, B or C possesses the faulty coin. Let $|A| = |B| = a$ and $|C| = n - 2a$. Observe that if the faulty coin is known to be in A then, at most, $f(a)$ further weighings are required; likewise, if the faulty coin is known to be in B. If the faulty coin is known to be in C then, at most, $f(n - 2a)$ further weighings will be required. It follows that the number of additional weighings required is, in any case, not more than $\max\{f(a), f(n-2a)\}$. In fact, we can be more precise: the number of additional weighings required in the worst case is *equal* to $\max\{f(a), f(n-2a)\}$. This implies that the total number of weighings needed (in the worst case) is $1 + \max\{f(a), f(n-2a)\}$.

To see how this helps to fix the values of f, consider the case $n = 4$. Clearly, $f(4) \geq 1$ and $f(4) \leq 2$ (for 2 weighings do indeed suffice; please check!); so $f(4) = 1$ or 2. For W_1, we must either do a 2-versus-2 or a 1-versus-1 weighing. Either way, the best that we can guarantee for ourselves, in the worst possible case, is the identification of a 2-element set which contains the faulty coin; this is so no matter how we do W_1. It follows from this that:

$$f(4) \geq 1 + \max\{f(1), f(2)\},$$

i.e., $f(4) \geq 2$. Since we already know that $f(4) \leq 2$, we conclude that $f(4) = 2$.

The nice thing about this argument is that it works for any value of n.

EXERCISE

11.1.2 Analyze the case $n = 5$, and show that $f(5) = 2$.

★ ★ ★

To proceed further, we need a symbol for the *ceiling function*.

• **Ceiling function** If x is a real number, $\lceil x \rceil$ denotes the least integer not less than x. For example,

$$\lceil 3.1 \rceil = 4, \quad \lceil 5 \rceil = 5, \quad \lceil 10.7 \rceil = 11, \quad \lceil -2.3 \rceil = -2, \quad \ldots.$$

EXERCISES

11.1.3 Prove that for any $n > 1$ we have $f(n+1) \geq f(n)$. (This looks obvious but it needs proof. One approach could be to start with an n-coin collection and assign a double identity or 'alias' to some coin.)

11.1.4 Prove that for any integer $n > 1$, we have

$$f(n) \geq 1 + f\left(\left\lceil \frac{n}{3} \right\rceil\right).$$

(First prove the following: If a, b, c are positive integers such that $a + b + c = n$, then $\max\{a, b, c\} \geq \lceil n/3 \rceil$.)

11.1.5 Prove that equality holds in the above problem; that is, for any integer $n > 1$,

$$f(n) = 1 + f\left(\left\lceil \frac{n}{3} \right\rceil\right).$$

(First show that $f(n) \leq 1 + f(\lceil n/3 \rceil)$.)

11.1.6 The relation in 11.1.5 can be used recursively to compute $f(n)$ for any $n > 1$.

EXAMPLE. Consider $n = 100$; since $\lceil 100/3 \rceil = 34$, $\lceil 34/3 \rceil = 12$, $\lceil 12/3 \rceil = 4$ and $\lceil 4/3 \rceil = 2$, we have, successively:

$$\begin{aligned} f(100) &= 1 + f(34), & f(34) &= 1 + f(12), \\ f(12) &= 1 + f(4), & f(4) &= 1 + f(2), \\ f(2) &= 1; \end{aligned}$$

therefore, $f(100) = 5$.

Compute $f(200)$ and $f(1000)$ using this technique.

11.1.7 Prove that $f(n) = \lceil \log_3 n \rceil$.

With this result, the analysis of the problem can be considered to be complete, for we have at hand not just a formula for $f(n)$ but, *implicitly contained in the formula*, the best possible weighing scheme.

11.1.8 Write down *explicitly* the optimal weighing scheme for the n-coin problem.

11.2 The counterfeit coin problem

We move to another version of the n-coin problem, which has now become a classic. As earlier, we are given an accurate 2-pan balance and n coins of identical appearance. Of these one is defective, but it is not known how its weight compares with that of a normal coin: *it could be heavier or lighter.* What is the least number of weighings needed to identify the defective coin and also the nature of its defect? (We assume, as always, a worst-case outcome.)

Let $g(n)$ be this number. Note that we must have $n > 2$: with $n = 2$ we make no progress no matter how many weighings are permitted! Thus, $g(n) = \infty$ for $n = 1, 2$.

EXAMPLE. Consider the case $n = 3$. Certainly $g(3) > 1$: a single weighing cannot possibly reveal the nature of the weight defect (even in a best-case scenario). On the other hand, two weighings suffice in all possible cases, so $g(3) \leq 2$. Therefore, $g(3) = 2$.

Notation Label the coins 1, 2, 3, ..., n. The action of weighing coins $\{1, 2\}$ against coins $\{3, 4\}$, with $\{1, 2\}$ in the left pan and

$\{3,4\}$ in the right pan, will be denoted thus:

$$\left(1,2 \;\middle|\; 3,4\right).$$

We summarize the result of each weighing via the use of the symbols L, E, R as follows:

- L if the left pan is heavier;
- E if the two pans have equal weight;
- R if the right pan is heavier.

If coin $\#k$ is faulty, then we write k^+ or k^- depending on whether the coin is heavier or lighter than a normal one. And by, say, the notation $\{2^+, 3^-\}$, we shall mean the following: *The weighings have revealed that one of the following possibilities is true: either coin #2 is counterfeit and heavier than normal, or coin #3 is counterfeit and lighter than normal.*

⋆ ⋆ ⋆

For $n = 3$, a possible weighing scheme is the following:

$$W_1 = (1 \mid 2), \qquad W_2 = (1 \mid 3).$$

The conclusions to be drawn are as indicated below. 'Imp' is short for '*impossible*'; the point is that certain combinations of outcomes are impossible.

Outcome	LL	LE	LR	EL	EE	ER	RL	RE	RR
Conclusion	1^+	2^-	Imp	3^-	Imp	3^+	Imp	2^+	1^-

Observe that we have listed 9 possible sequences of outcomes (each weighing gives rise to 3 possibilities, ∴ the number of possible outcomes is $3 \times 3 = 9$) and 6 possible deductions, each of which is associated with one sequence of outcomes. The assertion that $g(3) \leq 2$ follows from the table.

⋆ ⋆ ⋆

For the n-coin case, suppose that k weighings suffice to identify the faulty coin as well as the nature of its defect. Clearly, the weighing scheme must be such that every possibility is taken into account. There are 3^k possible sequences of outcomes, of which one, EE ... E, is useless. If the pans stay level each time, then we cannot possibly identify the weight status of the defective coin. (We might be able to identify the defective coin, but there is no way of identifying the nature of its defect.) Of the remaining $3^k - 1$ outcomes, some may be mutually contradictory, but it is difficult to say anything definite at the outset—it will depend on the weighing scheme. The set of possibilities for the defective coin is:

$$\{1^+, 1^-, 2^+, 2^-, \ldots, n^+, n^-\},$$

or $2n$ possibilities in all. Since the weighing scheme is such that each sequence of outcomes implies a definite conclusion, with no possibility left out, the number of possible outcomes cannot be less than the number of possible deductions. That is, we must have

$$3^k - 1 \geq 2n, \quad \therefore\ 3^k \geq 2n + 1, \quad \therefore\ k \geq \log_3(2n+1).$$

This yields a lower bound on $g(n)$, the least number of weighings required that will cover every possibility:

$$g(n) \geq \lceil \log_3(2n+1) \rceil.$$

For $n = 4$ this yields $g(4) \geq \lceil \log_3 9 \rceil = 2$; but we already know this! Can we make a tighter statement?

Consider the $n = 4$ case. We shall show that we cannot have $g(4) = 2$. Since only 4 coins are available, we can work only with sets of 2-against-2 or 1-against-1. There are three ways of weighing 2-against-2:

$$(1, 2 \mid 3, 4), \quad (1, 3 \mid 2, 4), \quad (1, 4 \mid 2, 3).$$

Suppose the two weighings are:

$$W_1 = (1, 2 \mid 3, 4), \qquad W_2 = (1, 3 \mid 2, 4).$$

Observe that coins #1 and #4 lie on opposite sides in both weighings. *This means that the scheme is unable to distinguish between* 1^+ *and* 4^- *and between* 1^- *and* 4^+. Likewise, the scheme

$$W_1 = (1, 2 \mid 3, 4), \qquad W_2 = (1, 4 \mid 2, 3)$$

cannot distinguish between 1^+ and 3^-, and between 1^- and 3^+. So these schemes will not work.

More generally, suppose that $W_1 = (a, b \mid c, d)$. Since we must be able to distinguish between the possibilities a^+ and c^- and between a^- and c^+, we cannot have a and c on opposite sides in W_2. The two coins cannot both be left out of W_2 either (this will not distinguish between a^+ and c^-, and between a^- and c^+). If only one of these two coins is used in W_2, then also there are pairs of possibilities between which we are unable to distinguish. For example, if $W_2 = (a \mid b)$ then we cannot distinguish between c^+ and d^+, and between c^- and d^-. So we are forced to place a and c together in one pan in W_2. This reasoning applies equally well to a and d: these two coins as well must be placed in one pan. Clearly, these two requirements are impossible to meet simultaneously. Similar reasoning may be used if $W_1 = (a \mid b)$. We conclude from this that 2 weighings will not, in general, suffice; that is, $g(4) \geq 3$. With this established, it is easy to show that $g(4) = 3$; we simply have to exhibit a scheme that works. Here is one such scheme (we shall show why it works):

$$W_1 = (1,2 \mid 3,4); \quad W_2 = (1,3 \mid 2,4); \quad W_3 = (1,4 \mid 2,3).$$

The possible outcomes are LLL, LLE, LLR, LEL, ..., or $3^3 = 27$ possibilities in all. However, the strings containing an E cannot occur, as all 4 coins are used in each weighing. This leaves $2^3 = 8$ possibilities. The possibilities with regard to which coin is the defective one, and in what manner, are also 8 in number. The two sets of possibilities may be matched against one another as shown below.

Result	LLL	LLR	LRL	LRR	RLL	RLR	RRL	RRR
Conclusion	1^+	4^-	3^-	2^+	2^-	3^+	4^+	1^-

The 1–1 correspondence shows that the scheme works, and it follows that $g(4) = 3$.

For $n = 6$, here is a scheme that does *not* work:

$$W_1 = (1,2 \mid 3,4), \quad W_2 = (1,3 \mid 5,6), \quad W_3 = (2,5 \mid 4,6).$$

Do you see why? (The scheme cannot distinguish between 2^+ and 4^-, nor between 2^- and 4^+.)

EXERCISES

11.2.1 Find a weighing scheme for $n = 5$ that does the needful in 3 weighings, and conclude that $g(5) = 3$.

11.2.2 Find a weighing scheme for $n = 6$ that does the needful in 3 weighings, and conclude that $g(6) = 3$.

11.2.3 Show that $g(n+1) \geq g(n)$ for each $n \geq 3$.

★ ★ ★

Query Observe that $g(n) = 3$ for $n = 4$, 5 and 6, and $g(n) \geq 4$ for $n \geq 13$. At which point does $g(n)$ jump from 3 to 4?

11.3 Solution to the 12-coin problem

In this section, we shall depart from our usual practice of 'feeling' our way into a problem and arriving at a solution through intuition and investigation; instead, we present a complete algorithm for the $n = 12$ case. The algorithm is sufficiently astonishing as to inspire incredulity. (It is due to Freeman J Dyson, a well-known mathematician, physicist and writer.)

Let the twelve coins be labelled 1, 2, 3, ..., 11, 12. We start by affixing an extra label to each coin, computed thus. For $n \in \{1, 2, 3, \ldots, 11, 12\}$, we denote by A_n and B_n respectively, the base-3 forms of n and $26 - n$, written as 3-digit numbers, with a 0 in front if necessary.

EXAMPLE. $A_5 = 012, \quad B_5 = 210, \quad A_6 = 020, \quad B_6 = 202.$

Observe that B_n is obtained from A_n by replacing the zeroes by twos and the twos by zeroes (the ones stay unchanged), and that $A_n \neq B_n$ for each $n \in \{1, 2, 3, \ldots, 12\}$. (If $A_n = B_n$, then we must have $A_n = 111$; but this corresponds to $n = 13$.)

Now, we categorize the labels A_n and B_n as 'clockwise' or 'anticlockwise' according to their *first change of digit, reading from left to right.* We regard the following changes of digit as 'clockwise'

$$0 \to 1, \quad 1 \to 2, \quad 2 \to 0;$$

and the following changes as 'anticlockwise':

$$0 \to 2, \quad 2 \to 1, \quad 1 \to 0.$$

EXAMPLE. $A_5 = 012$ is categorized as clockwise, as its first change of digit is from 0 to 1; while $B_5 = 210$ is categorized as anticlockwise, as its first change of digit is from 2 to 1.

Observe that for each n, one each of A_n and B_n is of the clockwise type; the other is of the anticlockwise type. We fix the label L_n of coin n thus:

$$L_n = \begin{cases} A_n & \text{if } A_n \text{ is of the clockwise type,} \\ B_n & \text{if } B_n \text{ is of the clockwise type.} \end{cases}$$

The following table displays the labels of the various coins.

n	L_n
1	001
2	220
3	010
4	011

n	L_n
5	012
6	202
7	201
8	200

n	L_n
9	122
10	121
11	120
12	112

Observe that each coin now has a clockwise-type label. The weighings now proceed as follows.

For W_1 we scan through the <u>first</u> digits of these labels. The coins for which this digit is 0 go into the *left* pan and those for which the digit is 2 go into the *right* pan. The weighing scheme thus starts with

$$W_1 = \quad (1,3,4,5 \mid 2,6,7,8).$$

For W_2, we consider the <u>second</u> digit of each label. The coins for which this digit is 0 go into the left pan and those for which the digit is 2 go into the right pan:

$$W_2 = \quad (1,6,7,8 \mid 2,9,10,11).$$

Finally, for W_3, the coins with 0 as the <u>third</u> digit of the label go into the left pan and those with a 2 go into the right pan:

$$W_3 = \quad (2,3,8,11 \mid 5,6,9,12).$$

Each weighing results in an outcome that is one of $\{\mathsf{L}, \mathsf{E}, \mathsf{R}\}$. From the three outcomes we produce a 3-digit signature label S, whose i^{th} digit is fixed by the outcome of W_i using the following code:

$$\mathsf{L} = 0, \quad \mathsf{E} = 1, \quad \mathsf{R} = 2.$$

EXAMPLE. The sequence of outcomes, LRE, produces the signature label $S = 021$, while the sequence, RRE, produces the signature label $S = 221$.

Obviously, S is either clockwise or anticlockwise. If it is the former, then $S = A_k$ for some k, or else $S = B_k$ for some k. This is so because the labels A and B exhaust between them all possible labels.

The identity of the culprit is now revealed: it is coin $\#k$, and the coin is *heavy* if $S = A_k$ and *light* if $S = B_k$.

EXAMPLE 1. Suppose that $S = 221$. Since $B_1 = 221$, we infer that coin #1 is the faulty coin and that it is lighter than normal. The reader should check that 1^- produces the sequence of outcomes RRE, whose signature label is $S = 221$. This is consistent with what was stated above.

EXAMPLE 2. Suppose that $S = 010$. Since $A_3 = 010$, we infer that the defective coin is coin #3 and that it is heavy. (The sequence of outcomes is LEL; please check!)

EXAMPLE 3. Suppose that $S = 021$. Since $B_7 = 201$, we infer that the defective coin is coin #7 and that it is light. (The sequence of outcomes is LRE.)

And that's the scheme! It is so very novel that it inspires many questions, and we shall pose them in the following sequence of problems.

COMMENT. The solution presented here was published in October 1946 by F J Dyson and R C Lyness, in the *Mathematical Gazette.* It is worth remarking that till it was discovered, it was thought that such problems could not be solved without 'branching' or 'back-tracking'.

EXERCISES

11.3.1 Why is it that $A_n \neq B_n$ for each n?

11.3.2 Why is it that in $\{A_n, B_n\}$ one number is of the clockwise type, while the other is of the anticlockwise type?

11.3.3 The weighing scheme is based on the labels of the coins: for W_k, we place in the left pan those coins for which the kth digit of the label is 0, and in the right pan those coins for which the kth digit is 2. Very conveniently, we find that each time there are four coins in each pan, with four coins left over. Is this an "accident", or could it have been foreseen?

11.3.4 The final step uses the clockwise/anticlockwise nature of the signature S to deduce the nature of the weight defect. Why does this work?

11.3.5 Generalize the scheme so that it can handle different numbers of coins. Show how the scheme can handle the case $n = 39$ and (more generally) $n = (3^k - 1)/2$ for $k > 1$. Use the algorithm to show that

$$g\left(\frac{3^k - 3}{2}\right) \leq k.$$

11.3.6 Prove that for $k > 1$,

$$g\left(\frac{3^k - 3}{2}\right) = k.$$

You will need to use the result of 11.3.5 together with an inequality obtained in the preceding section.

11.3.7 (The final decisive assertion.) Prove that $g(n) = k$ for precisely those values of n that are identified by the inequality:

$$\frac{3^{k-1} - 3}{2} < n \leq \frac{3^k - 3}{2}.$$

Example. $g(n) = 3$ for $n = 4, 5, 6, \ldots, 11, 12$, and for no other value of n.

11.3.8 Speculate on how Dyson may have hit upon this algorithm.

⋆ ⋆ ⋆

The condition given in Problem 7 may be written in the form

$$3^{k-1} < 2n + 3 \leq 3^k,$$

and from this we deduce the desired formula: $g(1) = g(2) = \infty$, and for $n > 2$,

$$g(n) = \left\lceil \log_3(2n + 3) \right\rceil.$$

Example. $g(10) = \lceil \log_3 23 \rceil = 3, \quad g(20) = \lceil \log_3 43 \rceil = 4.$

11.4 Miscellaneous problems

We close this chapter with a few miscellaneous coin-weighing problems.

11.4.1 You are given a set of 1001 coins of identical appearance, one of which has an 'X' written upon it. You are told that:

- 40 of the coins are defective, and the weight of a faulty coin differs from that of a good coin by exactly 1 gm;
- among the defective coins, coins of both types (heavy and light) are present.

You also have at hand an accurate pan balance, with a scale that displays the weight difference (in gms) between the contents of the two pans. How can the weight status of X be found in a *single* weighing?

11.4.2 We are given an accurate pan balance and two sets of coins, one marked A_1, A_2, A_3, A_4 and the other marked B_1, B_2, B_3, with weights w such that

$$w(A_1) < w(A_2) < w(A_3) < w(A_4),$$
$$w(B_1) < w(B_2) < w(B_3).$$

It is known that all the 7 coins have unequal weights. The problem is to arrange the coins in the order of increasing weight; that is, to weave the As with the Bs in the right order. What is the least number of weighings needed to accomplish this?

11.4.3 (Continuation) We are given an accurate pan balance and two sets of coins, one marked $A_1, A_2, \ldots, A_m$ and the other marked $B_1, B_2, \ldots B_n$. It is known that the $m+n$ coins all have unequal weights, and that the weights w obey the following conditions:

$$w(A_1) < w(A_2) < \cdots < w(A_m),$$
$$w(B_1) < w(B_2) < \cdots < w(B_n).$$

How many times does the balance need to be used in order to arrange the $m+n$ coins in order of increasing weight?

11.5 References

[1] Bagchi, B. "Bachet's Problem", in *Resonance*, November 1996: 88–91.

[2] Buchman, A L and Hawthorne, F S. Coin Weighings Revisited, in *Two-Year College Mathematics Readings*, MAA, 1981.

[3] Pranesachar, C R. "The Odd Ball Problem", in *Resonance*, March 1997: 76–78.

[4] Ball, W W Rouse. *Mathematical Recreations and Essays.*

[5] Pedoe, Dan. *The Gentle Art of Mathematics.*

Reference [1] is about a famous problem first posed by Bachet: *What set of weights will suffice to weigh every integral weight from 1 pound till 100 pounds, using a two-scale balance?* (The weights may be placed in either pan.) Reference [3] is about the counterfeit coin problem, considered more generally (for $n > 12$ coins).

Chapter 12

On Games of Strategy

12.1 Introduction

Game-playing clearly goes very far back in time[1], as excavations reveal (in Egypt, Greece, Mohenjodaro, etc.). Over the centuries, human beings have invented vast numbers of games, and most of us have some favourite game or other. To the problem solver, however, not all games are equal; some are far more rich and more interesting than others. Depending upon the particulars, some games may not have much to offer by way of 'theory', while others may have a great deal to offer; board games and games involving numbers are particularly rich in this regard. What distinguishes such games is the possibility of devising a precise strategy for winning. Broadly, one might refer to them as *games of strategy*. In this chapter, we shall discuss several such games.

⋆ ⋆ ⋆

COMMENT. There is scope for devising winning strategies in practically any 2-person game! (football, cricket, hockey, and so on). However, these strategies are typically heuristic in nature and of necessity imprecise; at best, they offer a greater or lesser probability of winning. It follows that such games also offer a lot of scope for developing intuition and expertise; hence, the need for a coach, a captain, a second-in-command, and so on. The games discussed here offer no such scope!

[1]Some thinkers go so far as to declare that free uncommitted play is one of the highest forms of human activity.

12.2 The ground rules

We shall confine our study to games with the following features:

- The game is between two players who play in alternation. They cannot abstain from making moves; that is, there is no 'passing'.
- The game is finite: it must end within a finite number of moves.
- There is no element of randomness in the game (∴ no rolling of dice, tossing of coins, etc.).
- Complete information is freely available to the players, and manuals may be referred to at any point in the game to find the best possible move. There is no possibility of hidden information; everything is above board—literally!

⋆ ⋆ ⋆

The twin requirements of finiteness and non-randomness rule out many of the familiar games: e.g., tennis, badminton, volleyball. Card games are ruled out by the requirement that complete information must be available to both players at every stage; likewise for games that depend upon dice throws or other randomizing devices.

Does chess pass the requirement of finiteness? This is not clear. It *is* possible for a game of chess to go on indefinitely; for instance, when each side is left with only the King and a Bishop. The usual convention in such cases is to agree to a draw, or for the arbiter to declare the game as drawn. Be that as it may, we shall not take up chess for study in this chapter, as it is far too deep and complex a game to be allotted just one chapter in a book—it would merit an entire book (of several volumes!) all to itself.

Here are some examples of games that do pass the tests.

- *The Twenty Questions Game–1*

  The first player selects a number n within some given range, say $1 \leq n \leq 10^6$. The second player's task is to find n by asking questions which must all be of the type that can be answered with a 'yes' or a 'no'. No more than 20 such questions are permitted. How should the second player proceed to find n?

- *The Twenty Questions Game–2*

  Here the first player selects some famous person, X, and the second player's task to identify X by asking questions, which must, as earlier, be of the type that can be answered with 'yes' or 'no' E.g, "Is X a man?", "Is X an Indian?", ... What strategy should the second player adopt?

- *Tic-Tac-Toe* (or *'Noughts-and-Crosses'*) This is a game known to every student and often played during school hours!

  The game is played upon a 3×3 grid. Playing alternately, the players place 'noughts' (O) and 'crosses' (X) in the cells of the grid—X for the first player and O for the second player. The first person to place three symbols of a kind in a line (row, column or diagonal) wins the game. Where should the first player place the first X, and how should the opponent respond? With the best possible play, can the first player force a win? (The answer to this is 'no': the game *can* end in a draw.)

- *Qubic*

  This is a 3-dimensional version of Tic-Tac-Toe—it is played on a $3 \times 3 \times 3$ cube, rather than a 3×3 grid. There are 27 cells available for placement of the Xs and Os. What is the result when the best possible strategies are followed?

  The game can also be played on a $4 \times 4 \times 4$ cube.

- *Digital Tic-Tac-Toe*

  As in Tic-Tac-Toe, this is played on a 3×3 grid. Playing in alternation, the players enter the digits 1, 2, 3, . . . , 8, 9 into the cells of the grid, one digit per cell. Once a digit is 'used' it cannot be used again. The first player uses the even digits, while the second player uses the odd digits. The first player to complete a line (row, column or diagonal) in such a way that the sum of the numbers in the line is 15 is the winner. What is the best strategy for the players to follow?

  If the first player is allowed to use the odd digits, she has an easy winning strategy. What is this strategy?

- *Nim*

  This is just one of the many 'take-away' games that enjoy great popularity. The game is played using heaps of matchsticks between two players, who play in alternation. Each move consists of selecting a heap and removing some matches from it. At least one matchstick must be removed, but this is the only restriction; the entire heap may be removed if so desired. The matches taken away have no further role to play in the problem and are set aside. The player who removes the *last* match is the winner. What is the best strategy?

⋆ ⋆ ⋆

It turns out that for the games that pass the tests listed above (finiteness, non-randomness, perfect availability of information), it is possible to conduct a complete analysis of the game with regard to the strategies available for play. In many of these cases, the mathematics involved in the analysis of the game is rich and elegant, as we shall see below.

12.3 Analysis of the games

Notation The first player will be called ***A***, and the second player ***B***. The task at hand is to analyze each game described below and to find the winning strategies.

12.3.1 Hundred coins

In front of the two players lies a heap of 100 coins. The players, in alternation, remove a desired number of coins from the heap—at least 1 coin and at most 7 coins. The player who removes the last coin wins. What is the winning strategy?

Here is another description of the game. Starting with an initial running total s of 0, the two players alternately add to s any number $n \in \{1, 2, 3, 4, 5, 6, 7\}$ of their choice. Thus, s is replaced by $s + n$; we denote this by writing $s \to s + n$. The objective is to be the first to reach a running total of exactly 100; whoever does so is declared the winner.

Example. ***A*** may start the game with a 5 (so $s = 5$), then ***B*** may add 6, giving $s = 11$, and ***A*** may now add 4, giving $s = 15$; and so on.

EXERCISE

12.3.1 With perfect play, who wins the game, ***A*** or ***B***?

⋆ ⋆ ⋆

More generally, suppose that at the start the heap contains N coins. The players remove coins in alternation, at least 1 and at most k coins on each move. The player who removes the last coin wins. What is the winning strategy? (The answer will, of course, depend on both k and N.)

12.3.2 Twenty questions

The first player ***A*** selects any integer n where $1 \leq n \leq 10^6$, and ***B*** is allowed to ask ***A*** up to twenty questions, each to be answered with a 'yes' or 'no'; ***B***'s objective is to find n. (We assume that ***A*** is honest!)

EXERCISES

12.3.2 Show that:

(a) it is possible for ***B*** to deduce ***A***'s number; but

(b) if the number of questions allowed is less than 20, then ***B***'s success cannot be guaranteed.

12.3.3 Here is a much more difficult version of the game. ***A*** selects any number n from the set $\{1, 2, 3, \ldots, 100\}$, and ***B*** is permitted to ask a maximum of m questions of ***A***, each to be answered with a yes or no. The difference is that ***A*** is permitted to lie *once* during the course of the game. (He may also choose not to lie at all!) At the end, ***B*** must know the number chosen by ***A***.

What is the least m that will guarantee success for ***B***?

12.3.3 Matchsticks

We start with a single heap containing N matchsticks (N is fixed in advance). ***A*** starts by dividing it into two *unequal* heaps, each containing at least 1 matchstick. (The number of heaps thus increases by 1.) ***B***, in turn, selects a heap and divides it into two unequal, non-empty heaps. This continues in alternation: at each stage a heap is selected and divided into two unequal heaps, each heap containing at least 1 matchstick. The game ends when a player is unable to accomplish the task prescribed; she is declared the loser.

EXERCISE

12.3.4 Make a complete analysis of the game. Determine, for each N, which player has a guaranteed win, and state the winning line of play.

EXAMPLE. Here is a possible game history for $n = 6$. The sizes of the heaps are indicated by the numbers. By the notation $6 \xrightarrow{A} (2,4)$ we mean that ***A*** has split a heap of size 6 into two heaps of sizes 2 and 4.

$$6 \xrightarrow{A} (2,4) \xrightarrow{B} (2,1,3) \xrightarrow{A} (2,1,1,2).$$

Since ***B*** is unable to move, ***A*** is declared the winner.

12.3.4 Chocolates

We start with a heap of some given number chocolates, say N. ***A*** starts by eating *no more than half of them*. ***B*** then eats no more than half the chocolates still remaining. The players proceed in this manner, alternately eating no more than half the chocolates still remaining in the heap. The game ends when a player is unable to accomplish the task prescribed; she is declared the loser.

EXERCISE

12.3.5 Make a complete analysis of the game. Determine, for each N, which player has a guaranteed win, and state the winning line of play.

EXAMPLE. Here is a possible game history for $N = 20$. The number of chocolates remaining in the heap at each stage is indicated as shown below. (Note that for each move of the type $a \rightarrow b$, we have $b \geq a/2$.)

$$20 \xrightarrow{A} 12 \xrightarrow{B} 7 \xrightarrow{A} 4 \xrightarrow{B} 2 \xrightarrow{A} 1.$$

Since ***B*** is unable to move, ***A*** is declared the winner.

12.3.5 Quadratic equations

The starting configuration here is a quadratic equation in x:

$$ax^2 + bx + c = 0,$$

in which the values of a, b, c have yet to be assigned.

The game proceeds as follows. ***A*** starts by assigning an integer value to one of a, b, c; then ***B*** assigns an integer value to one of the remaining coefficients, and then ***A*** assigns an integer value to the 3rd coefficient. We now have a quadratic equation in which each coefficient has been assigned an integer value, the only restriction being that a must be non-zero.

A wins if the equation has rational roots; otherwise ***B*** wins.

Exercise

12.3.6 Which player has a guaranteed win? What happens if the roles of 'rational' and 'irrational' are interchanged?

Make a complete analysis of the game.

12.3.6 Sprouts

In this section, we shall briefly discuss the game of Sprouts, invented by J H Conway and M S Paterson.

The game is played on *dots* (or 'points') and *edges* (segments that connect the dots); the edges may be line segments or portions of curves (the shape does not matter). The starting configuration is simply a set of dots, with no connecting edges, say something like this:

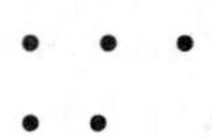

The starting number of dots, N, is any positive integer, and the starting number of edges is 0.

A move consists of: (a) joining a dot to itself by a loop and placing a new dot somewhere on the loop; or (b) joining two dots by an edge and placing a new dot somewhere on the edge. Either way, one new dot and two new edges are introduced. The following must be ensured: *the edges must not intersect one another.* Thus, at each stage of the game, we have a configuration of dots

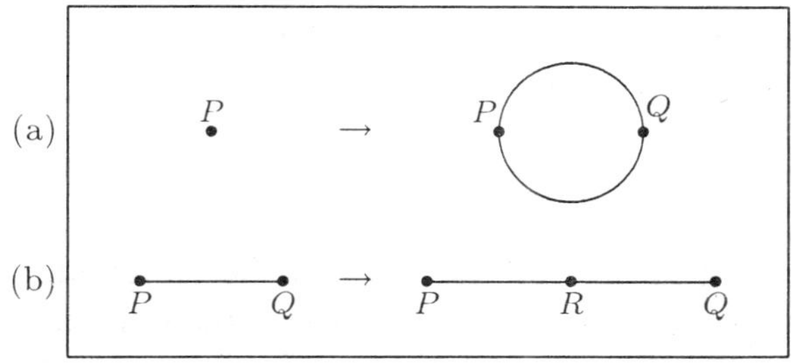

Figure 12.1. *The game of Sprouts*

and edges connecting some of them. The stopping condition is the following: *At no stage must any dot be connected to more than three other dots.* When it happens that no further edges can be drawn, the game is over; the player who is unable to do anything further is declared the loser.

The moves are illustrated in Figure 12.1. In (a) we have drawn an edge connecting dot P to itself (i.e., a loop), and in (b) we have drawn an edge connecting dots P and Q. In both cases, a new dot has been placed on the newly-drawn edge.

Observe that the stopping condition implies that action (a) can be done only if dot P is connected to no more than one other dot, and action (b) can be done only if dots P and Q are each joined to no more than two other dots.

Let the starting configuration have N dots and no connecting edges. Who wins the game with good play, the first player or the second player? We argue as follows.

In any intermediate configuration for any dot X, we refer to the number of dots connected to X as the *degree* of X. So the degree of a dot is simply the number of edges emanating from it. The stopping condition may now be stated succinctly: *The degree of a dot cannot exceed 3.*

With each intermediate configuration we associate a pair (n, e) giving the number of dots and edges in it (n is the number of dots and e the number of edges); we refer to (n, e) as its *state*. The starting state is $(N, 0)$, and since at each stage we add an extra dot and two new edges, the state changes in the following manner as the game progresses:

$$(N, 0) \to (N+1, 2) \to (N+2, 4) \to (N+3, 6) \to \cdots.$$

Clearly, after i moves the state of the game is $(N+i, 2i)$.

Let (N_f, e_f) denote the state of the game when it ends. Observe that the game can end in any of the following ways:

(a) every dot has a degree of 3 (every dot is 'full up'; no more edges can be drawn);

(b) every dot except one has a degree of 3, and there is exactly one dot with degree 2 ($\therefore$ the dot with a degree of 2 cannot be joined to any other dot);

(c) there *are* two or more dots with degree 2 or less available, but they have become isolated from one another because of the loops, so no more connecting edges can be drawn.

These are the only possibilities for the terminal position of the game. We shall show shortly that possibility (a) cannot happen (claim (1), below). For convenience, in the exposition we refer to (b) as an 'unblocked ending' and to (c) as a 'blocked ending'.

We now make the following claims.

(1) *In an unblocked ending, at the terminal position every dot except one has a degree of 3, and precisely one dot has a degree of 2.*

If there were a dot with a degree of 1, or two or more dots with degree of 2 or less, accessible to one another, then further loops or edges can be drawn; i.e., further moves are possible. So at the terminal stage in an unblocked ending, there is at most one dot with a degree of 2; every other dot has a degree of 3. However, it is not possible that there are no dots at all with a degree of 2, because after each move the newest or 'youngest' dot has a degree equal to 2. So, at the terminal stage, there must be one (and only one) dot with a degree equal to 2.

(2) *At the terminal stage in an unblocked ending, we have the following relation connecting the number of dots and the number of edges:* $e_f = (3N_f - 1)/2$.

This is so because from (1), in an unblocked ending, every dot but one has a degree of 3 and one dot has a degree of 2. If we consider all the dots in turn and make a running total of the edges that emanate from each dot, then the total we will reach is $3N_f - 1$. In this count, each edge is counted twice (once from "each end"); so we have $2e_f = 3N_f - 1$, i.e., $e_f = (3N_f - 1)/2$.

(3) *In an unblocked ending, the number of dots at the start and end are related by the equation* $N_f = 4N - 1$.

The state of the game changes in the following way:

$$(N,0) \to (N+1,2) \to (N+2,4) \to (N+3,6) \to \cdots.$$

So if the game has an unblocked ending and lasts a total of k moves, then $N_f = N + k$ and $e_f = 2k$. These relations, together with the one obtained above, yield: $k = N_f - N$, $\therefore e_f = 2(N_f - N)$, therefore,

$$2(N_f - N) = \frac{3N_f - 1}{2}, \qquad \therefore\ 4(N_f - N) = 3N_f - 1.$$

Solving this equation, we get $N_f = 4N - 1$, as claimed.

(4) *In an unblocked ending, a game lasts for precisely* $3N - 1$ *moves; and in a blocked ending, a game lasts for less than* $3N - 1$ *moves.*

For, from (3) we get $k = N_f - N = 3N - 1$. If the ending is blocked, then the game will have ended before these many moves are made.

⋆ ⋆ ⋆

The last result shows that the game cannot go on forever; the number of moves cannot exceed $3N-1$. If the ending is unblocked (no dots are isolated from one another), then the winner may be found as follows. If N is odd, then the game lasts an even number of moves, so the 2nd player (***B***) wins; and if N is even, then the game lasts an odd number of moves, so the 1st player (***A***) wins. It follows that if N is odd, then ***A*** will try to steer towards a blocked ending by isolating some dots; and if N is even, then it is ***B*** who will try to do this.

EXAMPLE. Consider the case $N = 2$. The game can last, at most, 5 moves. Since N is even, ***A*** will try for an unblocked ending, but ***B*** will attempt to isolate some dot so that the game ends in 4 moves. However, ***A*** can prevent this by drawing a loop from dot 1 to itself and placing dot 3 on the new edge. Now ***B*** is unable to isolate any dot; so ***A*** wins. (The reader should check this analysis.)

EXERCISE

12.3.7 Make a complete analysis of Sprouts for $N = 3$.

12.3.7 Nim

We come to the most celebrated and well-known of all take-away games: Nim. As described earlier, the game is played between two players who play alternately using heaps of matchsticks. Starting from an arbitrary configuration (i.e., various given heaps of matches), each move consists of selecting a heap and removing some matches from it. At least one matchstick must be removed, but this is the only restriction; the entire heap may be removed if so desired. The matches that are taken away have no further role to play in the problem and are set aside. The player who removes the *last* match is the winner.

Here is a sample game. We describe the state of the game at any point in the game, by giving the number of matchsticks in each heap. The starting configuration is $(2, 3, 4, 5)$.

$$\begin{aligned}(2,3,4,5) &\overset{A}{\to} (2,3,1,5) \overset{B}{\to} (2,1,5)\\ &\overset{A}{\to} (2,1,3) \overset{B}{\to} (2,1,1)\\ &\overset{A}{\to} (1,1) \overset{B}{\to} (1) \overset{A}{\to} (\,).\end{aligned}$$

Observe that ***A*** has won the game.

We shall describe a position in the game as *won for* A if it is ***A***'s turn to play and if, by means of a suitable move, he can force a win (this means that a win is guaranteed no matter how ***B*** responds). Similarly, it is *lost for* ***A*** if it is ***A***'s turn to play and he loses no matter what move he makes—assuming, of course, perfect play on the part of ***B***!

Exercises

12.3.8 Show that $(4, 5)$ is a win for ***A***. What should ***A*** do in order to force a win?

12.3.9 Show that (a, a) is a loss for ***A***.

12.3.10 Show that (a, b) is a win for ***A***, if a and b are unequal.

12.3.11 For which values of a is $(a, a + 1, 1)$ a win for ***A***; and, for these values, what is the winning move?

Hint. Focus on the parity of a.

12.3.12 Make a complete game analysis for the configuration $(a, a+2, 2)$.

12.3.13 Show that $(a, a+3, 3)$ is a win for ***A***, if and only if a is not a multiple of 4.

12.3.14 Make complete analyses for the positions $(a, a+1, 2)$ and $(a, a+2, 1)$.

12.3.15 Classify all pairs of positive integers (a, b) for which the position $(a, b, 1)$ is a win for ***A***.

12.3.16 Conduct a similar analysis for the position $(a, b, 2)$.

⋆ ⋆ ⋆

Readers who have persevered with these problems will, by now, have acquired a good intuitive feel for the game. They will doubtless be pleasantly surprised to find that there exists a complete mathematical theory behind Nim, one that enables any position to be classified as won or lost and indicates (in the former case) how a win can be forced.

It turns out that the theory is a very general one; with small modifications, it can be applied to many games of this type. We shall elaborate on this comment in a later section.

12.4 A mathematical approach to Nim

We start by making some general observations about any 2-person game of the type we are studying—games that are finite and non-random, and which have perfect availability of information.

Imagine a listing of *all* possible positions of the game, and suppose that the list can be partitioned into sub-lists **I** and **II**, with the following four characteristics:

(S1) Any move from a type **I** position necessarily leads to a type **II** position.

(S2) From any given type **II** position, it is possible to move to a type **I** position; there exists at least one move that will accomplish this.

(S3) The final position of the game, from which no further moves can be made, is of type **I**.

(S4) There exists an easy-to-compute criterion by which the type (**I** or **II**) of a position may be determined, and (in case it is of type **II**) to find a move that leads to a type **I** position.

If these conditions apply, the conclusion is clear: *the player-to-move wins from any type* **II** *position but loses from any type* **I** *position.* (We assume that both the players know the classification criteria!—recall that information is freely available.)

It is easy to see why. Given a type **II** position, the player-to-move (say ***A***) chooses a move that leads to a type **I** position. (By assumption S2 this is always possible.) The next move (***B***'s) necessarily leads back to a type **II** position (assumption S1), so the *status quo* is restored. This continues: ***A*** is repeatedly presented with a type **II** position and ***B*** with a type **I** position. Ultimately, we reach a position from which no further moves are possible, because the game cannot go on forever. By assumption, this position is of type **I**, and since it is ***B***'s turn to move, ***A*** is the winner.

From a human point of view, such a game would be of interest only if 'most' positions are of type **II**; that is, if a randomly chosen starting position is more likely to be of type **II** than type **I**. In this case, the odds are heavily in favor of the first player.

EXAMPLE. Consider the 'Hundred Coins' game described earlier (Section 12.3.1). The type **I**/type **II** classification for this game is easy to make. Let s be the number of coins still left at any intermediate stage. Then we classify a position as type **I** if $s \equiv 4 \pmod 8$, and as type **II** if $s \not\equiv 4 \pmod 8$.

Since $100 \equiv 4 \pmod 8$, the conditions listed above hold. For example, S1 holds because the number to be added lies between 1 and 7, and S3 holds because $100 \equiv 4 \pmod 8$. It follows that ***A*** can win by removing 4 coins (as his first move). For subsequent moves, he chooses the 8s complement of whatever ***B*** chooses; i.e., *if* ***B*** *removes x coins, then* ***A*** *removes $8 - x$ coins.*

⋆ ⋆ ⋆

It turns out that for Nim too there exists a type **I**/type **II** classification. The criterion is quite an intricate one, and is computed as follows. We write the heap sizes as binary numbers, in columnar format as in an addition problem, and then compute the sum of these numbers using the following addition rules:

$$0 + 0 = 0, \quad 0 + 1 = 1 = 1 + 0, \quad 1 + 1 = 0.$$

Readers may recognize this as 'exclusive-or addition'; in computer jargon, 'XOR' addition. The sum is called the NIM-sum. We give two examples below.

EXAMPLE 1. For the position $(3, 4, 5)$, with binary equivalents of 11 for 3, 100 for 4 and 101 for 5, we write:

$$\begin{array}{ccc} & 1 & 1 \\ 1 & 0 & 0 \\ 1 & 0 & 1 \\ \hline 0 & 1 & 0 \\ \hline \end{array}$$

In the bottom row we write down the XOR sum of the numbers 11, 100 and 101; the answer is 010. So the NIM-sum is 010.

EXAMPLE 2. For the position $(2, 3, 6, 7)$, we write:

$$\begin{array}{ccc} & 1 & 0 \\ & 1 & 1 \\ 1 & 1 & 0 \\ 1 & 1 & 1 \\ \hline 0 & 0 & 0 \\ \hline \end{array}$$

The NIM-sum, in this case, is $10 + 11 + 11 + 111 = 000$.

The classification scheme The criterion used is the following.

> *A position is classified as type* **I** *if every digit in the NIM-sum is 0, or else it is classified as type* **II**.

Thus, $(3, 4, 5)$ is of type **II** because of the 1 in 010, whereas $(2, 3, 6, 7)$ is of type **I** because each digit in 000 is 0.

The terminal position $(0, \ldots, 0)$ of a Nim game is clearly of type **I**, so condition S3 holds; and the criterion given above is easy-to-compute, so condition S4 holds. It remains to show that conditions S1 and S2 also hold. We illustrate S2 by showing how to move from the type **II** position $(3, 4, 5)$ to a type **I** position. Here is the array for the $(3, 4, 5)$ position:

$$\begin{array}{ccc} & 1 & 1 \\ 1 & 0 & 0 \\ 1 & 0 & 1 \\ \hline 0 & 1 & 0 \\ \hline \end{array}$$

Focusing on the 1 in the NIM-sum (010), we see that if the 11 in the first row were to be replaced by 01, then the NIM-sum would become 000 and our objective would be fulfilled. So the answer is: *move to* $(1, 4, 5)$. (It so happens that in this example, this is the only such move available.)

Here is another example. The position $(1, 3, 7, 9)$, with a NIM-sum of 1100, is of type **II**:

$$\begin{array}{cccc} & & & 1 \\ & & 1 & 1 \\ & 1 & 1 & 1 \\ 1 & 0 & 0 & 1 \\ \hline 1 & 1 & 0 & 0 \\ \hline \end{array}$$

We see that if the last row were to be replaced by 0101, then the NIM-sum would change to 0000. The correct move is therefore: *remove 4 matches from the largest heap; that is, move to the position* $(1, 3, 7, 5)$:

$$\begin{array}{ccc} & & 1 \\ & 1 & 1 \\ 1 & 1 & 1 \\ 1 & 0 & 1 \\ \hline 0 & 0 & 0 \\ \hline \end{array}$$

We have succeeded in moving to a type **I** position.

⋆ ⋆ ⋆

The tasks before us now are the following. *(1) Show that given any position whose NIM-sum has only 0s, any move whatever results in a position whose NIM-sum has at least one 1; and (2) given a position whose NIM-sum has one or more 1s, there exists at least one move to a position whose NIM-sum has only 0s*. These tasks we relegate to the problem set below.

Exercises

12.4.1 Complete the analysis of Nim by showing that conditions S1 and S2 hold.

12.4.2 Invent variations of Nim, and examine whether there exist natural classification schemes that apply to these versions of the game.

Here is a possible generalization: *the players are allowed to pick up matchsticks from any* <u>*two*</u> *heaps.*

12.5 References

[1] Averbach, B & Chern, O. *Problem Solving Through Recreational Mathematics*, Freeman, 1980.

[2] Ball, W W Rouse. *Mathematical Recreations and Essays*, Macmillan, 1960.

[3] Conway, John Horton. *On Numbers and Games*, Academic Press, 1977.

[4] Hardy, G H and Wright, E M. *An Introduction to the Theory of Numbers*, Cambridge University Press.

[1] and [2] are highly recommended. [3] has a very enigmatic definition of the notion of *number* which is reminiscent of the definition of real numbers via Dedekind cuts (encountered in the axiomatic construction of the real number system). [4] has a discussion on the game of Nim; the classification criterion which we presented above is proved here.

Chapter 13

Dessert!

These is no 'theory' in this chapter: no prelude, no preamble; only problems and, following them, more problems. Tackle them, pit your wits against them; and, yes, don't forget to enjoy yourself . . . !

EXERCISES

13.1.1 Tolerant fractions Consider the fraction 16/64. Noting the '6' on both sides of the fraction bar, we cancel it and obtain 1/4; and now (surprise!):

$$\frac{16}{64} = \frac{1}{4}; \quad \text{the cancellation has been correct!}$$

Find many more instances of this sort. Even better, find all such instances.

13.1.2 Cooperative square roots Consider the following equation:

$$\sqrt{2\,\frac{2}{3}} = 2\sqrt{\frac{2}{3}}.$$

(Check that the equation is correct.) Find many more instances of this sort. Even better, find all such instances.

13.1.3 Search for instances of the above that involve cube roots.

13.1.4 Find the least positive integer n with the property that $n/2$ is a square, $n/3$ is a cube, $n/5$ is a fifth power, and $n/7$ is a seventh power.

13.1.5 The figure below shows a rectangle subdivided into 4 rectangles, with areas as indicated. Find x in terms of a, b, c.

a	b
c	x

13.1.6 (Continuation) The figure below shows a rectangle subdivided into 9 rectangles by lines drawn parallel to its sides. The areas of 6 of these rectangles are as indicated in the figure. Find x in terms of a, b, c, d, e.

	a	b
c	d	
e		x

13.1.7 (Continuation) Redo the above two problems with the word *area* replaced by *perimeter*.

13.1.8 One hundred chess players enter a chess tournament organized on a Swiss-league system. The pairings are made randomly in each round, and each 'odd man out' is given a straight entry into the next round (a 'bye'). Losers are eliminated from the tournament after each round. (We assume that there are no draws.) Find out:

(a) the total number of games played;

(b) the number of rounds needed to complete the tournament.

13.1.9 (Continuation) Redo Problem 13.1.8 in more general terms, with '100' replaced by n (i.e., n players enter the tournament, $n > 1$).

13.1.10 (Continuation: This is a more difficult version of the problem, with altered rules.)

Pairings continue to be made randomly, with each 'odd man out' given a 'bye', but *two* losses have to be accumulated for a player to be eliminated from the tournament. (As earlier, there are no draws.) Find, in terms of the number of players:

(a) the total number of games played;

(b) the number of rounds needed to complete the tournament.

13.1.11 Suppose that a and b are real numbers such that

$$a^2 + b^2 = 2, \quad a^3 + b^3 = 2, \quad a + b \neq 2.$$

What are the possible values of $a + b$?

13.1.12 Prove that: $\sqrt[3]{2+\sqrt{5}} + \sqrt[3]{2-\sqrt{5}} = 1.$

13.1.13 Solve for x:

$$\sqrt[3]{6x+28} - \sqrt[3]{6x-28} = 2.$$

13.1.14 A class has 25 children seated in a rectangular 5×5 array. The teacher announces that the children must reseat themselves as follows: *each student must move to a desk immediately to his right, left, front or back* (thus, diagonal moves are not permitted). Note that children seated in the corners have fewer desks to choose from than those more centrally placed. Is the change possible?

13.1.15 (Continuation) Suppose that the class had 16 children (instead of 25), seated in a square 4×4 array. Show that it is possible to reseat the children in accordance with the given rule. In how many ways can it be done?

13.1.16 For positive integers n written in base-10, let $\mathrm{Sum}(n)$ denote the sum of the digits of n.

EXAMPLE. $\mathrm{Sum}(47) = 11$, $\mathrm{Sum}(163) = 10$.

Find all positive integers n, for which

(a) $n = 2\,\mathrm{Sum}(n)$;

(b) $n = 3\,\mathrm{Sum}(n)$;

(c) $n = 7\,\mathrm{Sum}(n)$;

(d) $n = 30\,\mathrm{Sum}(n)$.

13.1.17 Using the 9 non-zero digits each just once, form two base-10 numbers A and B whose product is *as large as possible.*

13.1.18 Using the 9 non-zero digits once each, form two base-10 numbers A and B such that $A = 2B$.

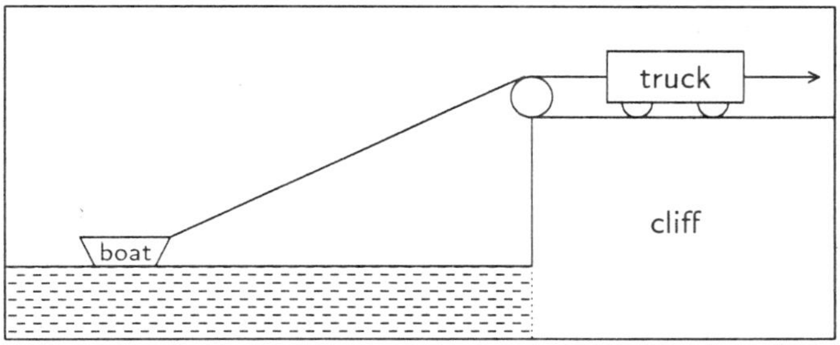

Figure 13.1. *Boat on a lake*

13.1.19 Let a be a digit and n a positive integer expressed in base-10 form. Can the pair (n, a) have the following property? *The number n' produced by inserting an a between the last two digits (the units and tens digits) of n is a multiple of n.*

13.1.20 Let m, n be two integers with $\gcd(m, n) = 5$. What possible values can $\gcd(m + n, m - n)$ take?

13.1.21 Consider the set of fractions of the form $\dfrac{12n + 7}{7n + 4}$:

$$\frac{7}{4}, \quad \frac{19}{11}, \quad \frac{31}{18}, \quad \frac{43}{25}, \quad \frac{55}{32}, \quad \frac{67}{39}, \quad \ldots.$$

Observe that the fractions shown are all in their lowest terms; the numerator and denominator are coprime. Is this true for all such fractions?

13.1.22 A boat lies moored upon a lake; it is attached by a taut rope to a truck (the rope passes over a pulley at the edge of the cliff) as shown in Figure 13.1. The truck moves away from the cliff by a distance of 5 meters. Let d be the distance moved by the boat. Is d greater or less than 5?

13.1.23 A cross made from five 1×1 squares is to be cut into smaller pieces, which are then to be reassembled into a square. How can this be accomplished? (No overlaps are allowed during reassembly, nor any gaps.) What is the least number of pieces into which the cross must be cut?

13.1.24 The blank spaces in the number below are to be filled with digits so that the resulting number is a multiple of 99:

$$2143\_62701\_4632.$$

In how many ways can this be done?

13.1.25 Sixteen chess players take part in a chess tournament, organized on a 'round-robin' system (every two players play precisely one game against one another). A win earns 1 point, a draw $\frac{1}{2}$ point, a loss 0 points (draws can happen). A player accumulating 10 or more points is awarded a grandmaster norm.

(a) What is the largest possible number of grandmaster norms that can be awarded at the end of the tournament? What is the least possible such number?

(b) Analyze the corresponding n-player problem.

13.1.26 Given a 4-digit number, I reverse its digits to form a new 4-digit number; then I compute the difference between the two numbers. What is the largest possible difference I can get?

13.1.27 Some unit cubes are stacked together to make a larger cube. Then, some faces of the larger cube are painted red. When the paint has dried, the larger cube is taken apart. It is found that 45 of the unit cubes have no paint on any of their faces. Can you tell how many faces of the larger cube were painted?

What happens when '45' is replaced by '100'?

13.1.28 Could the number $33\star\star6$ be a perfect square? (The stars represent missing digits. The two missing digits do not have to be the same.)

13.1.29 Find a 5-digit number which uses the digits 1, 2, 3, 4, 5 once each, and which is a multiple of 4 but not a multiple of 6.

13.1.30 The number $\star679\star$ is a multiple of 72. How many different ways are there of filling in the two missing digits?

13.1.31 Once, three travellers stopped at a restaurant for dinner; it cost Rs 30/-, so they paid Rs 10/- each. As they were leaving, they were stopped by the waiter who said that there had been a mistake in the bill, and that the cost actually was only Rs 25/-; so he returned Rs 5/- to them. Unable to see how to divide Rs 5/- in a proper way among three men, they gave Rs 2/- to the waiter as a tip and kept Rs 1/- each. So actually they paid Rs 9/- each for the dinner. Now $9+9+9+2=29$, but they had originally paid Rs 30/- in all. What has happened to the missing Rs 1/-?

13.1.32 Four points are to be marked on a flat surface so that there are only two different distances between them. For instance, the four points could be placed at the vertices of a square. You can check that there are now only two different distances between them.

There are five other ways of solving this problem. Find them!

13.1.33 Two cubical wire frames are given, one bigger than the other. Is it possible that the large cube can be passed through the small one?

13.1.34 Express as many numbers as you can from 1 to 100 using exactly four 4s and any mathematical symbol you know.

For example, 15 and 16 can be written as:

$$15 = (4 \times 4) - \frac{4}{4}, \quad 16 = (4 \times 4) + 4 - 4.$$

(There are other solutions for both the numbers.)

You are allowed to use any of the following signs: $+$, $-$, $\times$, $\div$, $\sqrt{}$, the 'dot' symbol for a recurring decimal (e.g., $.\dot{4}$ means the infinite decimal $0.44444\ldots$), and the exclamation mark (the symbol for factorials); e.g., 4! means $1 \times 2 \times 3 \times 4$ or 24.

You should be able to find at least one way of writing most, if not all, of the numbers from 1 to 100.

13.1.35 The result of dividing a 2-digit number with another 2-digit number with a calculator is as shown: 0.4482758. What were the two numbers?

13.1.36 The product of some three consecutive odd numbers is 357627. What are the three numbers?

13.1.37 Playing with a calculator a girl found that:

$$159 \times 48 = 7632,$$

and then noticed that the equation contained each of the digits 1, 2, ..., 8, 9 once only. She could hardly believe her luck and felt that the result must be unique.

But she was wrong! There are other ways in which this can happen. Can you find any of them?

13.1.38 With a 7-minute sand timer and an 11-minute sand timer what is the easiest way to time the boiling of an egg for 15 minutes?

13.1.39 You have 10 boxes and 44 marbles. Can you put your marbles in the boxes so that each box contains a different number of marbles?

13.1.40 While driving on a highway at constant speed, I pass a milepost with a 2-digit number. Exactly one hour later, I pass a milepost with the same two digits, but they are reversed. In another hour I pass a 3rd milepost, with the same two digits reversed and separated by a zero. What was the number on the 1st milepost?

13.1.41 Two candles of the same length are made of different materials so that one burns out completely at a uniform rate in 3 hours and the other in 4 hours. At what time should the candles be lit so that at 4 : 00 pm one is twice the length of the other?

13.1.42 A man plans on arriving by train at the station at 3:00 pm and to have his wife drive him home. Instead, he catches an earlier train and arrives at the station at 2:00 pm. He decides to walk towards home and is picked up by his punctual wife *en route*. They arrive home 20 minutes earlier than planned. How long did he walk?

13.1.43 Four men attended a ballet with their wives. In their hurry to leave, each man mistakenly picked up another man's coat and the hat of yet another man and left. Jim took the coat belonging to the man whose hat was taken by Dan, and Dan's coat was taken by the person who took Jim's hat. Kevin took Peter's hat. Whose hat and whose coat had Jim and Dan taken?

13.1.44 Five professors, residing at five different cities, teach five different subjects. From the clues, determine the subject that each professor teaches and the place where he is residing.

(a) Mr A does not stay at either Bangalore or Lucknow. He teaches philosophy.

(b) Mr B does not stay at either Hyderabad or Lucknow. He teaches Maths.

(c) Mr D stays at Jaipur.

(d) Mr E does not stay either at Bangalore or at Delhi. He teaches Geography.

(e) Mr C does not take History and he stays at Delhi.

13.1.45 In 1932, I was as old as the last two digits of my birth year. My grandfather said that that applies to him also. How old are we?

13.1.46 Between midnight and noon, how many times are the minute hand and the hour hand of a clock at right angles to each other?

13.1.47 Two identical squares $ABCD$ and $PQRS$, each of size 10×10, are placed so that P lies at the centre of square $ABCD$. Can you find the area of the overlapping region?

13.1.48 Four identical circles are fitted into a square of side 1 (each circle touches two sides, and the fit is tight). Find the radius of the largest circle that will fit in the central hole.

13.1.49 Three identical circles are fitted into an equilateral triangle of side 1 (each circle touches two sides of the triangle, and the fit is tight). Find the radius of the largest circle that will fit in the central hole.

13.1.50 A certain 10-digit number has among its digits one 1, two 2s, three 3s, and four 4s. Could the number be a perfect square?

13.1.51 Are there any triples a, b, c for which $a + b + c = a \times b \times c$? Can you find them all?

13.1.52 Arrange the digits 0, 1, 2, 3, ..., 7, 8, 9 into a 10-digit number such that the following property holds good: *the number formed by the 1st k digits of the number is divisible by k, for k* = *2, 3, 4, 5, 6, 7, 8, 9 and 10.*

13.1.53 $ABCDE$ is a pentagon with all its sides equal in length, and its angles satisfy the relation $A \geq B \geq C \geq D \geq E$. Must the pentagon be regular?

13.1.54 Let $ABCDE$ be a regular pentagon. Consider the ratio of the area of the pentagon to that of the 'pentagram' or five-pointed star, $ACEBD$. Is the ratio less than, equal to, or greater than 2?

13.1.55 What is the area of the largest equilateral triangle that can be inscribed in a square?

13.1.56 Place three different positive integers on the three vertices of a triangle so that the integers at the ends of each edge add up to a perfect square. What is the simplest possible solution?

13.1.57 Place four different positive integers on the four vertices of a square so that the integers at the ends of each edge add up to a perfect square.

13.1.58 A 9×12 rectangular piece of paper is folded so that two opposite vertices coincide. What is the length of the crease?

13.1.59 Show how to cut a square into exactly 11 squares. In how many different ways can this be done?

13.1.60 Can a square cake be cut into triangular pieces each of which is acute-angled?

13.1.61 Can a cube be sliced so that the cross-section is an equilateral triangle or a regular hexagon?

13.1.62 Can a regular tetrahedron be sliced so that the cross-section is a square?

13.1.63 I go for a walk one day between 4 pm and 5 pm, and return between 5 pm and 6 pm; I find that the hour- and minute-hand of my watch have changed places, exactly. For how long have I been away?

13.1.64 Are the second-hand, minute-hand and hour-hand of a clock ever inclined at exactly $120°$ to one another?

13.1.65 Located inside an equilateral triangle ABC is a point P such that $PA = 6$, $PB = 8$ and $PC = 10$. Find the area of the triangle.

13.1.66 What is the remainder when x^{100} is divided by $x^2 - 3x + 2$?

13.1.67 Find the number of sets X with the property that $\{1, 2\} \subset X \subset \{1, 2, 3, 4, 5\}$.

13.1.68 Inscribed in a circle is a quadrilateral with sides 25, 39, 52, and 60 (taken consecutively). What is the diameter of the circle?

13.1.69 Points C, A, M, B lie on a circle, in that order, and M is the midpoint of arc CAB. P is the foot of the perpendicular from M to chord AB. Given that $CA = 3$ and $AP = 5$, find the length of PB.

13.1.70 A mathematics teacher is rowing a boat upstream. As he passes under a low bridge, his hat is knocked off. Being quite absent minded, like all mathematics teachers, he notices only 5 minutes later that his hat is missing. He immediately turns around and rows downstream with equal effort. He retrieves his hat exactly one km from the bridge. What is the speed of the current?

13.1.71 Three circles, C_1, C_2, and C_3, are mutually tangent to one another, externally; C_1 and C_2 touch at Z, C_2 and C_3 touch at X, and C_1 and C_3 touch at Y. The lines XY, YZ and ZX are drawn and extended till they meet the remaining circle. Show that the two points of intersection thus created on each circle are the end-points of a diameter.

13.1.72 Write the integers 1, 2, 3, 4, 5, 6, 7 in a row, and then beneath them the same integers but in some other order. For example, we may get the following arrangement.

$$\begin{array}{ccccccc} 1 & 2 & 3 & 4 & 5 & 6 & 7 \\ 2 & 6 & 3 & 5 & 7 & 1 & 4 \end{array}$$

Now subtract one row from the other, taking only the absolute difference in each case (the sign does not matter); we get:

$$\begin{array}{ccccccc} 1 & 2 & 3 & 4 & 5 & 6 & 7 \\ 2 & 6 & 3 & 5 & 7 & 1 & 4 \\ \hline 1 & 4 & 0 & 1 & 2 & 5 & 3 \end{array}$$

Observe that 1 occurs twice in the last row. Show that no matter what the arrangement of the numbers in the 2nd row is, there will always be some number which occurs two or more times in the 3rd row.

13.1.73 Find all the integers between 200 and 300 that have exactly six positive divisors.

13.1.74 Mr M never lies except except on Sundays when he always lies. On which day or days of the week can he say, "If I didn't lie yesterday, then I will lie tomorrow"?

13.1.75 What is the least positive integer which has exactly 10 divisors?

13.1.76 How many non-congruent triangles are there in which each side has integer length less than or equal to 5?

13.1.77 Find x, y, z given the equations:

$$\begin{aligned} (x+y)(x+y+z) &= 66, \\ (y+z)(x+y+z) &= 99, \\ (z+x)(x+y+z) &= 77. \end{aligned}$$

13.1.78 The mid-point of an edge of a cube is connected by segments to the two corners furthest from it; what is the area of the resulting triangle? (Take the edge length of the cube to be 1.)

13.1.79 In order to list all the integers from 0 till 7 in binary form:

$$0,\ 1,\ 10,\ 11,\ 100,\ 101,\ 110,\ 111,$$

we need twelve 1s. (Count!) How many 1s would we need if we had to list all the integers from 0 till 1023 in binary form?

13.1.80 If $a+b=c$, $b+c=d$, $c+d=a$, and b is a positive integer, find the largest possible value of $a+b+c+d$.

13.1.81 The sum of the squares of three consecutive positive odd integers A, B, C is a 4-digit number N all of whose digits are the same. Find A, B, C, N.

13.1.82 Solve the following equation for x:

$$\left(x^2+x+1\right)\left(x^6+x^3+1\right)=\frac{10}{x-1}.$$

13.1.83 Let P be the centre of the square constructed on the hypotenuse AB of a right angled triangle ABC. Show that CP bisects angle ACB.

13.1.84 Show without any calculations that a triangle with sides 5, 5, 6 has the same area as a triangle with sides 5, 5, 8. Generate more such pairs of incongruent isosceles triangles with integer sides.

13.1.85 Show that there are infinitely many sets of five consecutive positive integers a, b, c, d, e such that $a+b+c+d+e$ is the cube of an integer and $b+c+d$ is the square of an integer.

13.1.86 Within a circle with centre O and radius 2 cm lie two smaller circles whose radius is 1 cm; they touch one another at O. A fourth circle is drawn which touches all three circles. Find the radius of this circle.

13.1.87 Find a number such that if its first digit (the left-most one) is deleted, the resulting number is $1/57$ of the original number.

13.1.88 Find a formula for the sum of the angles of a n-pointed star.

(A star is constructed as follows. Let a convex n-sided polygon be given. For $i = 1, 2, \ldots$ let the ith and $(i+2)^{\text{th}}$ sides of the polygon be extended till they meet. We assume that the two sides are non-parallel and do meet. These extended sides form the sides of the n-star.)

13.1.89 If three of the roots of the equation $x^4 + ax^2 + bx + c = 0$ are 1, 2 and 3, find the value of $a + c$.

13.1.90 Given that $0.9 < x < 1$, arrange x, y, z in increasing order, where

$$y = x^x, \quad z = x^{x^x}.$$

13.1.91 Find the number of solutions of the equation:

$$\sin x = \frac{x}{100}.$$

13.1.92 **A problem in set theory** We use the symbol Δ to denote 'symmetric difference'; if A and B are two sets, then $A \,\Delta\, B$ denotes their symmetric difference; that is, $A \,\Delta\, B = (A - B) \cup (B - A)$.

Let S be a set, and let A, B denote subsets of S. Let X stand for an unknown subset of S. Show that the equations (a) $A \cup X = B$ (b) $A \cap X = B$ do not necessarily have any solution for the unknown set X; and if they do have solutions, then in general there are multiple solutions.

On the other hand, the equation $A \,\Delta\, X = B$ always has *precisely one* solution, no matter what A and B are. Prove this statement.

13.1.93 A cube has three of its vertices at the points $(2, 3, 0)$, $(0, 5, 4)$ and $(4, 1, 8)$. Find the coordinates of the centre of the cube.

13.1.94 The lengths of the sides of a right-angled triangle are three consecutive terms of an AP. Show that the lengths are in the ratios $3 : 4 : 5$.

13.1.95 The tetrahedron $OABC$ is such that OA, OB, OC are mutually perpendicular. Prove that triangle ABC is not right angled.

13.1.96 A hexagon inscribed in a circle has three sides of length a and three sides of length b. Find the radius of the circle in terms of a and b.

13.1.97 Let n be a positive integer. Show that:

$$1 - \frac{1}{2} + \frac{1}{3} - \cdots + \frac{1}{2n-1} = \frac{1}{n} + \frac{1}{n+1} + \cdots + \frac{1}{2n-1}.$$

13.1.98 Show that the number of ways of selecting three numbers from the set $\{1, 2, \ldots, 3n\}$ such that their sum is a multiple of 3 is equal to

$$\frac{n(3n^2 - 3n + 2)}{2}.$$

13.1.99 The number 30! ends in 7 zeroes. What is the digit that precedes these zeroes?

13.1.100 The remainder theorem states: *Given a polynomial $f(x)$ and a real number a; then the linear form $x - a$ is a divisor of $f(x)$ if and only if $f(a) = 0$.*

Find a continuation of the following: *Given a polynomial $f(x)$ and a real number a; then the quadratic form $(x - a)^2$ is a divisor of $f(x)$ if and only if* __________.

Appendix A

Ascent and Descent

A.1 The method of induction

In this appendix, we shall briefly describe two extremely important proof techniques—*proof by induction*, which may also be called 'proof by ascent', and *proof by descent*. They will be found to be of use in solving some of the problems in this book. We first define the term 'proposition'.

• A *proposition* is simply a statement or claim; it asserts something, and it could be true or false. Thus, "$10 > 7$" and "$2 > 3$" are propositions (the first one is true, the second is false); "$2^n > n^2$ for all integers $n \geq 5$" is another proposition (true); and so is "the number $2^{2^n} + 1$ is prime for all positive integers n" (false).

Suppose that a given proposition is to be proved for all positive integers n; say, something like:

> *For all positive integers* n*, the sum* $1 + 2 + \cdots + n$ *equals* $n(n + 1)/2$.

There are several ways to prove this statement, but here is how it is done via induction.

Let $\mathcal{P}(n)$ denote the proposition "$1+2+\cdots+n = n(n+1)/2$". Then $\mathcal{P}(1)$ is true, for it asserts that $1 = 1$, which is true; so is $\mathcal{P}(2)$, which asserts that $1+2 = \frac{1}{2}(2 \times 3)$ or $3 = 3$. So $\mathcal{P}(n)$ is true for $n = 1$ and $n = 2$.

We now assume that $\mathcal{P}(k)$ is true for *some* positive integer k. This means that:

$$1 + 2 + \cdots + k = \frac{k(k+1)}{2} \quad \text{for some } k.$$

Adding $k+1$ to both sides of this equality, on the left side we get $1+2+\cdots+k+(k+1)$, and on the right side we get:

$$\frac{k(k+1)}{2}+(k+1)=(k+1)\cdot\left(\frac{k}{2}+1\right)=\frac{(k+1)(k+2)}{2},$$

after simplification. This means that:

$$1+2+\cdots+k+(k+1)=\frac{(k+1)(k+2)}{2},$$

which is just the proposition $\mathcal{P}(k+1)$. *So if $\mathcal{P}(k)$ is true, then so is* $\mathcal{P}(k+1)$. But we already know that $\mathcal{P}(1)$ is true; $\therefore$ $\mathcal{P}(2)$ is true; $\therefore$ $\mathcal{P}(3)$ is true; $\therefore$ $\mathcal{P}(4)$ is true; and so on, *ad infinitum*. That is, $\mathcal{P}(n)$ is true for all positive integers n, as had to be proved.

This is the 'method of induction'. It is of use in situations where, on the basis of observations ('mathematical experiments') we have arrived at some hypothesis. The hypothesis may be more a piece of guesswork than anything else. For instance, we may examine the statements:

$$1+2=2^2-1, \quad 1+2+2^2=2^3-1, \quad 1+2+2^2+2^3=2^4-1,$$

and be led to the guess that $1+2+2^2+\cdots+2^{n-1}=2^n-1$ for *all* positive integers n.

In general, suppose that we have a proposition $\mathcal{P}(n)$ whose truth we wish to establish; here $\mathcal{P}(n)$ is a proposition concerning an arbitrary positive integer n. Suppose that the following two conditions hold:

(a) The proposition $\mathcal{P}(1)$ is true (and readily verified).

(b) Given that $\mathcal{P}(n)$ is true, we may also show $\mathcal{P}(n+1)$ to be true; i.e., we have the logical implication:

truth of $\mathcal{P}(n)$ implies truth of $\mathcal{P}(n+1)$.

In this scenario, we may confidently assert that $\mathcal{P}(n)$ is true for all positive integers n. This is so because the truth of $\mathcal{P}(1)$ implies that of $\mathcal{P}(2)$, which implies the truth of $\mathcal{P}(3)$, which in turn implies the truth of $\mathcal{P}(4)$, and so on, *ad infinitum*.

In writing out a proof based on induction, the verification of steps (a) and (b) must be shown clearly. Step (a) is referred to as *anchoring* the induction, and (b) is the *inductive step*.

Sometimes we may need to anchor the induction at a number greater than 1; for instance, in showing that $2^n \geq n^2$ for integers $n \geq 5$. In this case, the steps given above are replaced by: (a) verify that $\mathcal{P}(k)$ is true for some positive integer k; and (b) show that for $n \geq k$ the truth of $\mathcal{P}(n)$ implies the truth of $\mathcal{P}(n+1)$. With these steps done, the truth of $\mathcal{P}(n)$ is established for all integers $n \geq k$.

Solved examples We now present some examples showing the use of induction; they may be treated as 'case studies'.

EXAMPLE 1. Show that for all positive integers n, we have:

$$1^2 + 2^2 + 3^2 + \cdots + n^2 = \frac{n(n+1)(2n+1)}{6}.$$

SOLUTION. Let $\mathcal{P}(n)$ denote the assertion to be proved. Then $\mathcal{P}(1)$ is true (it asserts that $1^2 = (1 \cdot 2 \cdot 3)/6$ or $1 = 1$); and so is $\mathcal{P}(2)$, which asserts that $1^2 + 2^2 = (2 \cdot 3 \cdot 5)/6$ or $5 = 5$. So the induction has been anchored.

Now assume that $\mathcal{P}(k)$ is true for some positive integer k; then

$$1^2 + 2^2 + 3^2 + \cdots + k^2 = \frac{k(k+1)(2k+1)}{6}.$$

Adding $(k+1)^2$ to both sides, we get $1^2 + 2^2 + 3^2 + \cdots + (k+1)^2$ on the left, and

$$\begin{aligned}\frac{k(k+1)(2k+1)}{6} + (k+1)^2 &= (k+1) \cdot \left(\frac{2k^2+k}{6} + (k+1)\right)\\ &= (k+1) \cdot \frac{2k^2+7k+6}{6}\\ &= \frac{(k+1)(k+2)(2k+3)}{6}\end{aligned}$$

on the right. The form of the expression shows that the truth of $\mathcal{P}(k)$ implies that of $\mathcal{P}(k+1)$. As the anchoring has been done, the truth of $\mathcal{P}(n)$ for integers $n \geq 1$ is established. □

EXAMPLE 2. Show that the quantity $n(n+1)(n+2)$ is a multiple of 6 for all positive integers n.

SOLUTION. Let $f(n) = n(n+1)(n+2)$. Then $f(1) = 6$, which is a multiple of 6. So the stated proposition is true for $n = 1$; the induction has been anchored. Next, observe that:

$$\begin{aligned} f(n+1) - f(n) &= (n+1)(n+2)(n+3) - n(n+1)(n+2) \\ &= 3(n+1)(n+2). \end{aligned}$$

So the difference between $f(n+1)$ and $f(n)$ is a multiple of 3. Indeed, it is a multiple of 6, because one out of $n+1$ and $n+2$ is even, implying that the product $(n+1)(n+2)$ is even. This means that $f(n+1) = f(n) +$ some multiple of 6. Therefore, if $f(n)$ is a multiple of 6, then so is $f(n+1)$. We have now succeeded in carrying out the inductive step. The proof is now complete; we conclude that $n(n+1)(n+2)$ is a multiple of 6 for all positive integers n. □

EXAMPLE 3. Show, via induction, that $2^n > n^2$ for integers $n \geq 5$.

SOLUTION. Let $\mathcal{P}(n)$ denote the statement that $2^n > n^2$. Observe that $\mathcal{P}(n)$ is false for $n = 1, 2, 3$ and 4, but is true for $n = 5$. So we anchor the induction at $n = 5$.

Now suppose that $2^k > k^2$ for some integer $k \geq 5$; then we have $2^k > k^2$, $\therefore\ 2^{k+1} > 2k^2$. But $2k^2 > (k+1)^2$ for all integers $k \geq 3$; for this inequality is equivalent to $k^2 - 2k > 1$, or $(k-1)^2 > 2$, or $k > \sqrt{2} + 1$, i.e., $k \geq 3$ since k is an integer. So if $k \geq 5$ then certainly we have $2k^2 > (k+1)^2$, and so we get $2^{k+1} > (k+1)^2$. This proves the inductive step, and the result is established. □

EXERCISES

A.1.1 Show that 9 is a divisor of $2^{2n} - 3n - 1$ for all integers $n > 0$.

A.1.2 Show that that for all integers $n > 0$,

$$\frac{1}{3} + \frac{1}{15} + \frac{1}{35} + \cdots + \frac{1}{(2n)^2 - 1} = \frac{n}{2n+1}.$$

(The denominators are $2^2 - 1 = 3$, $4^2 - 1 = 15$, $6^2 - 1 = 35$, and so on.)

A.1.3 Show, via induction, that $2^n > n^3$ for integers $n \geq 10$.

A.1.4 Show, using induction, that the sum of the interior angles of a convex n-sided polygon is $(n-2) \times 180^\circ$.

A.1.5 Show, via induction, that $a^{2n} - b^{2n}$ has $a + b$ as a factor.

A.2 The method of descent

In contrast to the principle of ascent described above, we have the *principle of descent*, which states simply that *a non-empty subset of the natural numbers* $\mathbf{N}$ *has a least member*. The following corollary emerges naturally from this principle:

> *Let A be a subset of $\mathbf{N}$ with the property that if $k \in A$ then $k-1 \in A$; in other words, every element in A has an immediate predecessor (in A). Then A is the empty set.*

The reason is clear. If A is non-empty, then it has a least member, say n. The defining property ("every element in A has an immediate predecessor") implies that $n-1 \in A$; but this contradicts the fact that n is the least member of A. Therefore A is empty.

Readers will readily appreciate that the principle of descent is logically the same as the principle of induction. *Each of these principles implies the other.*

Solved examples We now present some examples showing how the principle of descent is put to use; they may be treated as 'case studies'.

Example 1. Show that $\sqrt{2}$ is irrational.

Solution. We have encountered this proposition earlier, but now we present a proof based on descent. The strategy will be to assume that the claim is false, and then use descent to show that a contradiction has taken place.

Suppose that $\sqrt{2}$ is rational. Then there exist positive integers n such that $n\sqrt{2}$ is an integer. Let A be the set of such positive integers (i.e., $A = \{n \ : \ n\sqrt{2} \in \mathbf{N}\}$), and let a be the *least* integer in A.

Consider the quantity $b = a\sqrt{2} - a = a(\sqrt{2}-1)$. It is an integer for the simple reason that $a\sqrt{2}$ and a are integers, and since $0 < \sqrt{2} - 1 < 1$ we get $0 < b < a$. So b is a positive integer which is less than a. Now observe that:

$$b\sqrt{2} = (a\sqrt{2} - a) \times \sqrt{2} = 2a - a\sqrt{2},$$

showing that $b\sqrt{2}$ is an integer (for it is the difference of two integers). So $b \in A$, and b is smaller than a. This contradicts the assumption made at the very beginning that a is the *least* integer in A. We conclude that A has no least member; in other words, that A is empty. This implies that $\sqrt{2}$ is irrational.

EXAMPLE 2. Show that the only solution in integers of the equation

$$x^2 + y^2 + z^2 = 2xyz$$

is $x = 0,\ y = 0,\ z = 0$.

SOLUTION. We shall show, more generally, that the only solution in integers of the equation:

$$x^2 + y^2 + z^2 = \text{a multiple of } 2xyz \tag{2.1}$$

is $(x, y, z) = (0, 0, 0)$.

Suppose that the given equation possesses a solution in positive integers. Let (x_1, y_1, z_1) be the solution for which the sum $x_1 + y_1 + z_1$ is the least possible. Since the quantity $2x_1y_1z_1$ is even, it follows that x_1, y_1, z_1 are either all even <u>or</u> that two of them are odd and one of them is even. If the latter happens, assume (with no loss of generality) that x_1 is even, whereas y_1, z_1 are odd. This leads to:

$$\begin{aligned} &x_1^2 \equiv 0 \pmod 4, \quad y_1^2 \equiv z_1^2 \equiv 1 \pmod 4, \\ &2x_1y_1z_1 \equiv 0 \pmod 4, \end{aligned}$$

an absurdity, for it leads to $x_1^2 + y_1^2 + z_1^2 \equiv 2 \pmod 4$ whereas $2x_1y_1z_1 \equiv 0 \pmod 4$. Therefore, x_1, y_1, z_1 are all even; say

$$x_1 = 2x_2, \quad y_1 = 2y_2, \quad z_1 = 2z_2 \quad (\text{with } x_2, y_2, z_2 \in \mathbf{N}).$$

Substituting these in the equation $x_1^2 + y_1^2 + z_1^2 =$ some integer times $2x_1y_1z_1$, we obtain:

$$\begin{aligned} &4x_2^2 + 4y_2^2 + 4z_2^2 = 8 \times \text{an integer} \times 2x_2y_2z_2, \\ \therefore \quad &x_2^2 + y_2^2 + z_2^2 = \text{an integer} \times 2x_2y_2z_2. \end{aligned}$$

Since the sum $x_2 + y_2 + z_2 = (x_1 + y_1 + z_1)/2 < x_1 + y_1 + z_1$, we see that the existence of a solution (x_1, y_1, z_1) in positive integers of equation (2.1) implies the existence of another solution in positive integers with half the sum. This step can be iterated indefinitely, leading to an absurdity, and we are forced to conclude that the equation has no solutions in non-zero integers.

EXERCISES

A.2.1 Show, using the descent principle, that the quantity $\sqrt{3}$ is irrational.

A.2.2 Show, using the descent principle, that the 'golden ratio':

$$\phi = \frac{\sqrt{5}+1}{2}$$

is irrational.

HINT. Use the following property of ϕ:

$$\frac{\phi}{1} = \frac{1}{\phi - 1}.$$

A.2.3 (USAMO, 1976 Problem 3)
Find all integral solutions of the equation:

$$a^2 + b^2 + c^2 = a^2 b^2.$$

A.2.4 Find all integral solutions of the equation:

$$a^2 + b^2 + c^2 + d^2 = 2abcd.$$

A.3 References

[1] Sury, B. "Mathematical Induction—an Impresario of the Infinite", in *Resonance*, February 1998: 69–76.

Appendix B

Glossary of Terms

- *Arithmetic mean (average).* The arithmetic mean of two numbers a, b is the quantity $\frac{a+b}{2}$.
- *Arithmetic progression.* A sequence of numbers $a_1, a_2, a_3, a_4, \ldots$ is said to be in arithmetic progression, or AP for short, if

$$a_2 - a_1 = a_3 - a_2 = a_4 - a_3 = \cdots .$$

The quantity $d = a_2 - a_1$ is called the *common difference* of the AP.
- *Binomial coefficient.* Let n, k be integers, with $0 \le k \le n$; by $\binom{n}{k}$ we mean the quantity

$$\frac{n!}{k!(n-k)!}.$$

It is the coefficient of x^k in the expansion of $(1+x)^n$. The coefficient is defined for other values of n too; i.e., $n \notin \mathbf{N}$ (but k must be a non-negative integer):

$$\binom{n}{k} = \frac{n}{1} \times \frac{n-1}{2} \times \frac{n-2}{3} \times \cdots \times \frac{n-k+1}{k}.$$

It is now the coefficient of x^k in the power series expansion of $(1+x)^n$ in powers of x.
- *Binomial theorem.* This is the statement that if n is a non-negative integer, then

$$(1+x)^n = \binom{n}{0} + \binom{n}{1}x + \binom{n}{2}x^2 + \binom{n}{3}x^3 + \cdots + \binom{n}{n}x^n.$$

If $n \notin \mathbf{N}$, then the expansion of $(1+x)^n$ in ascending powers of x is an infinite series, and we have:

$$(1+x)^n = 1 + \frac{n}{1}x + \frac{n}{1}\frac{n-1}{2}x^2 + \frac{n}{1}\frac{n-1}{2}\frac{n-2}{3}x^3 + \cdots.$$

This expansion is valid only for $-1 < x < 1$.

- *Centroid of a triangle.* This is the point of intersection of the three medians of the triangle.
- *Ceva's theorem.* With reference to a given triangle ABC, this theorem states that for three points P, Q and R located on the sidelines BC, CA and AB (respectively) of triangle ABC, the lines AP, BQ and CR are concurrent if and only if the following relation holds:

$$\frac{BC}{PC} \cdot \frac{CQ}{QA} \cdot \frac{AR}{RB} = 1.$$

Here, the lengths BP, $PC, \ldots, RB$ are *signed*; they could be negative or positive.

- *Circumcentre of a triangle.* With reference to a given triangle ABC, its circumcentre is the centre of the circle that passes through A, B, C.
- *Circumcircle of a triangle.* With reference to a given ABC, its circumcircle is the circle that passes through A, B, C.
- *Composition of functions.* If, f, g are functions such that the range of g lies within the domain of f, then the *composite function* $f \circ g$ is defined thus: $f \circ g(x)) = f(g(x))$. That is, g acts first on the input x, and then f acts on the result of the first action.
- *Congruence (number theoretic).* If a, b, c are numbers, with c not equal to 0, then by the statement $a \equiv b \pmod{c}$, read as "a is congruent to b modulo c", we mean that $a - b$ is a multiple of c; that is, a and b leave the same remainder when divided by c.
- *Convex function.* A function f defined on an interval I is said to be convex if for all $a, b \in I$ the following relation holds:

$$f\left(\frac{a+b}{2}\right) \le \frac{f(a)+f(b)}{2}.$$

Examples of convex functions: $x \mapsto x^2, x \mapsto e^x$.
If the inequality is reversed then the function is said to be *concave*.

- *Cyclic polygon.* This refers to a polygon whose vertices lie on a circle; so the polygon can be *inscribed* in a circle.

- *de Moivre's theorem.* This is the statement that for all angles x and all integers n (positive or negative),

$$(\cos x + i \sin x)^n = \cos nx + i \sin nx.$$

If n is not integral, then the RHS is one of the values of the LHS. (Here $i = \sqrt{-1}$ is the imaginary unit.)

- *Determinant of a* 2×2 *matrix.* The determinant of the matrix $\begin{pmatrix} a & b \\ c & d \end{pmatrix}$ is defined to be the quantity $ad - bc$.

- *Dot product of two vectors.* Given two vectors $\boldsymbol{a}$ and $\boldsymbol{b}$, their scalar product or dot product, $\boldsymbol{a} \cdot \boldsymbol{b}$ is defined to be the scalar quantity $|\boldsymbol{a}||\boldsymbol{b}|\cos\theta$, where θ is the angle between the two vectors. (Here $|\boldsymbol{a}|$ and $|\boldsymbol{b}|$ refer to the lengths of the vectors $\boldsymbol{a}$ and $\boldsymbol{b}$.)
  Of particular significance is the fact that the dot product is zero if the two vectors are perpendicular to one another.
- *Escribed circle, excircle.* Given a triangle, there are precisely four circles that can be drawn in the plane of the triangle, touching the three lines on which its sides lie. Of these, one lies inside the triangle; this is the 'incircle'; the other three are the 'excircles' or 'escribed circles'.
- *Euclid's* GCD *algorithm.* Given two non-zero integers a, b, their GCD can be found by applying Euclid's GCD algorithm. This consists of repeatedly applying the mapping

$$(a, b) \mapsto (b, c), \quad \text{where} \quad c = \text{Rem}(a \div b),$$

till the 2nd coordinate becomes 0; once this stage is reached, the GCD is given by the 1st coordinate.
For example, if $a = 60$ and $b = 16$, we get:

$$(60, 16) \mapsto (16, 12) \mapsto (12, 4) \mapsto (4, 0).$$

So the GCD is 4 in this case.
- *Fermat's 'little' theorem.* This states that for all integers n and all prime numbers p, the quantity $n^p - n$ is a multiple of p; that is,

$$n^p - n \equiv 0 \pmod{p}.$$

COROLLARY. If n is an integer not divisible by p, then $n^{p-1} \equiv 1 \pmod{p}$.

- *Geometric mean.* If a, b are non-negative numbers, then their geometric mean is $\sqrt{ab}$.
- *Geometric progression.* A sequence of numbers $a_1, a_2, a_3, a_4, \ldots$ is said to be in geometric progression or GP is

$$\frac{a_2}{a_1} = \frac{a_3}{a_2} = \frac{a_4}{a_3} = \cdots.$$

The quantity a_2/a_1 is the *common ratio* of the GP.
- *Harmonic mean.* If a, b are non-zero numbers, then their harmonic mean is the quantity $\dfrac{2ab}{a+b}$.
More generally, the harmonic mean of a collection of non-zero numbers is the reciprocal of the arithmetic mean (AM) of their reciprocals.
- *Harmonic progression.* A sequence of non-zero numbers, $a_1, a_2, a_3, \ldots$ are said to be in harmonic progression, or HP, if their reciprocals $1/a_1, 1/a_2, 1/a_3, \ldots$ are in AP.
- *Heron's formula for area of a triangle.* This states that for $\triangle ABC$, with sides a, b, c and semi-perimeter $s = (a+b+c)/2$, its area $[ABC]$ is given by

$$[ABC] = \sqrt{s(s-a)(s-b)(s-c)}.$$

- *Homothety.* Two shapes are said to be homothetic to one another if one of them can be mapped onto the other, point for point, by an enlargement about some point by some suitable scale factor.
- *Incentre of a triangle.* With reference to a given triangle ABC, the centre of the circle that fits snugly inside the triangle, touching all the sides, is its incentre. It is the meeting point of the internal bisectors of the angles of the triangle.
- *Incircle of a triangle.* With reference to a given a triangle ABC, the circle that fits snug inside the triangle, touching all the sides, is its incircle.
- *Inverse function.* If f and g are functions such that $f \circ g = g \circ f =$ Id, where Id is the identity function (Id$(x)= x$ for all x), then f and g are inverses of one another. We write: $f^{-1} = g$, $g^{-1} = f$. For example, $x \mapsto e^x$ and $x \mapsto \ln x$ are a pair of inverse functions.
- *Law of triangles: cosine law.* This is the statement that in $\triangle ABC$ we have

$$c^2 = a^2 + b^2 - 2\,ab\,\cos C.$$

- *Law of triangles: sine law.* This is the statement that in $\triangle ABC$ we have
$$\frac{a}{\sin A} = \frac{b}{\sin B} = \frac{c}{\sin C} = 2R,$$
where R is the circumradius of the triangle.
- *Menelaus' theorem.* With reference to a given triangle ABC, this theorem states that three points P, Q and R located on the side-lines BC, CA and AB (respectively) of triangle ABC are collinear if and only if the following relation holds:
$$\frac{BP}{PC} \cdot \frac{CQ}{QA} \cdot \frac{AR}{RB} = -1.$$
Here the lengths BP, PC, ..., RB are signed; they could be negative or positive.
- *Orthocentre of a triangle.* This is the point of intersection of the three altitudes of the triangle.
- *Periodic function.* A function f from **R** into **R** is said to be periodic if for some non-zero real number a we have the identity $f(x+a) = f(x)$ for all $x \in$ **R**. The *period* of the function is the least value of a for which such an identity holds ('least' in absolute value). For example, the period of $x \mapsto \sin 2x$ is π.
- *Radical axis of two circles.* Given two circles C and C', the locus of points P such that the tangents from P to C and C' have equal length is termed the radical axis of the two circles.
- *Roots of unity.* A complex number w is a 'root of unity' if $w^n = 1$ for some positive integer n.
- *Vector.* A vector is an entity with a direction as well as a magnitude; it is defined by these two attributes. The best example of a vector is a displacement.
- *Zero of a function.* Given a function f acting on **R**, if $f(a) = 0$ for some $a \in$ **R**, then a is called a "zero" of f.

Appendix C

List of References

[1] Averbach, B and Chein, O. *Problem Solving Through Recreational Mathematics*, Freeman.

[2] Baderia, G S. *Fascination of Mathematical Puzzles*

[3] Ball, W W Rouse. *Mathematical Recreations and Essays*, Macmillan, 1960.

[4] Beiler, A H. *Recreations in the Theory of Numbers*, Dover, 1964.

[5] Bolt, Brian. *The Amazing Mathematical Amusement Arcade*, Cambridge University Press, 1984.

[6] Caroll, Lewis. *Mathematical Recreations of Lewis Caroll*, Dover.

[7] Conway, John. *On Numbers And Games*, Academic Press.

[8] Coxeter, H S M. *Introduction to Geometry*, Wiley, 1961.

[9] Coxeter, H S M, and Greitzer, S L. *Geometry Revisited*, MAA.

[10] Degrazia, Joseph. *Math Tricks, Brain Twisters & Puzzles*, Bell.

[11] Dudeney, Henry E. *Amusements in Mathematics*, Dover.

[12] Dudeney, Henry E. *Canterbury Puzzles*, Dover.

[13] Durell, C V. *Modern Pure Geometry*, Macmillan.

[14] Gardiner, A. *Mathematical Puzzling*, Oxford University Press, 1987.

[15] Gardner, Martin. *Mathematical Puzzles and Diversions*, Penguin.

[16] Gardner, Martin. *More Mathematical Puzzles and Diversions*, Penguin.

[17] Gardner, Martin. *Further Mathematical Diversions*, Penguin

[18] Golomb, S. *Polyominoes*, Scribner's, 1965

[19] Graham, L A. *Ingenious Mathematical Problems and Methods*, Dover, 1959.

[20] Hunter, J A. *Mathematical Brain Teasers*, Dover, 1976.

[21] Jacobs, Harold. *Mathematics: A Human Endeavour*, Freeman.

[22] Krantz, Steven. *Techniques of Problem Solving*, Universities Press.

[23] Larson, Loren C. *Problem-Solving Through Problems*, Springer Verlag, 1983.

[24] Leblanc, H, and Wisdom, W. *Deductive Logic*, Allyn and Bacon, 1976.

[25] Loyd, S. *Mathematical Puzzles of Sam Loyd*, Dover, 1959.

[26] Loyd, S. *More Mathematical Puzzles of Sam Loyd*, Dover, 1960.

[27] Mathematical Association of America. *Mathematics Magazine*, MAA.

[28] Mathematical Association of America. *College Mathematics Journal*, MAA.

[29] National Council of Teachers of Mathematics. *Mathematical Challenges II Plus Six*, NCTM, 1974.

[30] National Council of Teachers of Mathematics. *Mathematics Teacher*, NCTM.

[31] Newman, Donald J. *A Problem Seminar*, Springer Verlag, 1982.

[32] Niven, I, and Zuckerman, H. *An Introduction to the Theory of Numbers*, John Wiley, 1972.

[33] Ore, Oystein. *Theory of Graphs*, AMS.

[34] Polya, George. *Mathematical Discovery*, Wiley, 1981.

[35] Polya, George. *How to Solve It*, Princeton University Press, 1945.

[36] Salkind, C T and Earl, J M. *The Contest Problem Book III*, Mathematical Association of America, 1973.

[37] Smullyan, Raymond. *What is the Name of this Book?* Prentice-Hall, 1978.

Shirali, S A. *A Primer on Divisibility* , Universities Press, 2001.

Shirali, S A. *A Primer on Number sequences*, Universities Press, 2001.

[38] Smullyan, Raymond. *The Lady or the Tiger*, Knopf, 1982.

[39] Steinhaus, Hugo. *One Hundred Problems in Elementary Mathematics*, Basic Books, 1964

[40] Summers, G. *New Puzzles in Logical Deduction*, Dover, 1968.

[41] Summers, G. *Test Your Logic*, Dover, 1972.

[42] Trigg, C. *Mathematical Quickies*, Dover, 1967.

Index